Material Selection, Product Design and Additive Manufacturing Processes for Operational Sustainability

This book explores the integration of sustainability into the core of material selection, product design, and manufacturing processes, with a focus on additive manufacturing. It provides a comprehensive guide to operational sustainability, delving into environmentally conscious approaches that reduce waste, optimise resources, and enhance long-term operational efficiency. The scope covers the entire product lifecycle, from material selection to end-of-life recycling, offering insights into how emerging technologies contribute to sustainable development.

Features:

- Bridges the gap between theory and practice in sustainable material selection, product design, and additive manufacturing
- Highlights how additive manufacturing contributes to sustainability
- Describes generative design and topology optimisation
- Reviews the implementation of sustainable manufacturing processes
- Explores recycling, reuse, and reproduction design approaches

This book is intended for graduate students and researchers in metallurgical, mechanical, and manufacturing engineering, including materials science.

Advances in Design, Materials and Manufacturing for Sustainability
Series Editors: Jitendra Kumar Katiyar and T V V L N Rao

Conventional and Powder Mixed Electro-Discharge Machining
Biomedical Applications
Edited by Ahmad Majdi Abdul-Rani, Masdi Muhammad, T V V L N Rao, Saeed Rubaiee, Anas Ahmed and Mohd Danish

Micro- and Biofluidics
Structure, Properties, Technology and Spotlight into the Future
Avinash Kumar and Jitendra Kumar Katiyar

Natural Fiber Composites for a Sustainable Environment
Edited by Jibin Philip, Jerin J Jose and Soney George

Material Selection, Product Design and Additive Manufacturing Processes for Operational Sustainability
Ram Krishna

For more information about this series, please visit: www.routledge.com/Advances-in-Design-Materials-and-Manufacturing-for-Sustainability/book-series/CRCADMMS

Material Selection, Product Design and Additive Manufacturing Processes for Operational Sustainability

Ram Krishna

CRC Press
Taylor & Francis Group
Boca Raton London New York

CRC Press is an imprint of the
Taylor & Francis Group, an **informa** business

Designed cover image: © Ram Krishna

First edition published 2027
by CRC Press
2385 NW Executive Center Drive, Suite 320, Boca Raton FL 33431

and by CRC Press
4 Park Square, Milton Park, Abingdon, Oxon, OX14 4RN

CRC Press is an imprint of Taylor & Francis Group, LLC

ISBN: 9781032999173 (hbk)
ISBN: 9781041060994 (pbk)
ISBN: 9781003633754 (ebk)

DOI: 10.1201/9781003633754

Typeset in Times
by Newgen Publishing UK

Contents

Author Biography

Ram Krishna is an academic and researcher in the field of Metallurgical and Materials Engineering, with a strong focus on sustainable manufacturing and advanced materials processing. He is currently associated with the Department of Metallurgical and Materials Engineering at the National Institute of Technology (NIT) Jamshedpur, India.

His research interests include material selection for sustainability, additive manufacturing processes, product design optimisation, life cycle assessment, and the application of digital technologies such as Industry 4.0, IoT, and data-driven tools in manufacturing systems. He has contributed to research in areas related to sustainable materials, energy-efficient manufacturing, and circular economy practices, with an emphasis on practical industrial relevance.

Krishna has been actively involved in academic teaching, research supervision, and industry-linked projects. His work aims to bridge the gap between theoretical sustainability concepts and their real-world implementation in manufacturing industries. Through this book, he brings together technical knowledge, case-based insights, and operational perspectives to support students, researchers, and industry professionals working toward sustainable manufacturing practices.

Preface

Sustainability has become a central concern in modern manufacturing. Industries are expected to reduce environmental impact, use resources responsibly, and remain economically competitive at the same time. Achieving this balance requires changes not only in technology but also in everyday operational decisions.

This book focuses on operational sustainability through three key pillars: material selection, product design, and additive manufacturing processes. These areas strongly influence energy use, waste generation, product lifespan, and recyclability. Decisions made at these stages determine the overall sustainability performance of a product.

The book adopts a life cycle perspective, covering stages from raw material selection to end-of-life management. It highlights the role of tools such as life cycle assessment, design optimisation, and digital technologies in supporting sustainable outcomes. Additive manufacturing is discussed not as an inherently sustainable solution, but as a process whose sustainability depends on informed planning, process control, and material use.

The content is written for undergraduate and postgraduate students, researchers, and industry professionals. Emphasis is placed on clarity, practical relevance, and real-world application. The aim is to provide readers with a structured understanding of how sustainability can be implemented and measured within manufacturing operations.

Dr. Ram Krishna
Assistant Professor
Department of Metallurgical and Materials Engineering
NIT, Jamshedpur

Foreword

The proposed book is entitled 'Material Selection, Product Design, and Additive Manufacturing Processes for Operational Sustainability'. The term 'a sustainable future' refers to a state in which economic, social, and environmental sustainability are achieved in a balanced way, and it requires the use of materials and manufacturing processes that minimise environmental impact and conserve resources. This includes selecting materials that are renewable, biodegradable, and non-toxic, such as bamboo, hemp, and cork, as well as manufacturing processes that are energy-efficient and produce minimal waste, such as 3D printing. Additionally, it is important to consider the entire life cycle of a product, from raw material extraction to disposal, to ensure that it has the least possible impact on the environment by implementing closed-loop systems, in which resources are recycled, reused, or repurposed. The use of digital technologies such as life cycle assessment, Industry 4.0, industrial IoT, and sustainable design can also aid in optimising the product design and production and reducing environmental impact. Sustainable certifications and labelling schemes, such as Cradle to Cradle, are incorporated to ensure that products are environmentally friendly. Overall, the selection of materials and manufacturing processes is a crucial component in design for a circular economy and in creating a sustainable future.

Manufacturing industries across the world are under increasing pressure to reduce environmental impact while maintaining productivity and economic viability. In this context, operational sustainability has become a key requirement rather than an optional goal. This book addresses sustainability from a practical engineering perspective. It focuses on decisions that engineers and designers make every day – material choice, product design, and manufacturing process selection. By connecting these decisions to sustainability outcomes, the book provides a clear and realistic framework for implementation.

The strength of this work lies in its balanced treatment of theory and practice. It does not treat sustainability as an abstract idea, but as a measurable and manageable part of manufacturing operations. The inclusion of additive manufacturing, digital tools, and real industrial examples makes the content highly relevant to modern manufacturing environments.

This book will serve as a valuable reference for students, researchers, and professionals who wish to integrate sustainability into manufacturing systems in a structured and effective manner.

Vinod Kumar
General Manager – Federal
Phillips Machine Tools India Pvt. Ltd

Acknowledgements

The author would like to express sincere gratitude to colleagues, students, and industry professionals who contributed through discussions, feedback, and practical insights. Special thanks are extended to academic peers and industrial collaborators whose experiences helped shape the case studies and applied perspectives presented in this book.

The author also acknowledges the support of the National Institute of Technology Jamshedpur for providing an academic environment that encourages research and interdisciplinary collaboration. Appreciation is extended to reviewers and publishing professionals for their guidance during the preparation of this manuscript.

Finally, the author is grateful to family members for their continued support and encouragement throughout the development of this work.

1 Introduction

1.1 IMPORTANCE OF OPERATIONAL SUSTAINABILITY IN MATERIAL SELECTION, PRODUCT DESIGN, AND ADDITIVE MANUFACTURING PROCESSES

Integrating operational sustainability into material selection, product design, and additive manufacturing brings a paradigm shift in how industries manage their manufacturing lifecycle. Operational sustainability signifies monitoring and managing manufacturing processes in businesses in a way that can be sustained over the long term without harming environmental and socio-economic systems [1]. In practice, this concept aligns with the triple bottom line of sustainability – people, planet, and profit – ensuring that individual activities remain environmentally friendly, socially equitable, and economically viable [2, 3]. Figure 1.1 illustrates the Triple Bottom Line framework for operational sustainability in manufacturing.

Bringing sustainability principles to strategies minimises resource depletion, reduces waste generation, promotes safe working conditions, and contributes positively to societies. Further, the implementation brings robust monitoring to track process performance and identify the need for planning and improvement to ensure the modification of manufacturing cycles to achieve sustainability [4].

By regularly monitoring the performance indicators, industries can gain a vital understanding of the effectiveness of their sustainable manufacturing processes. This thorough analysis helps identify areas needing improvement, allowing companies to make informed decisions and implement the necessary corrective actions. Long-term operational sustainability and the inclusion of sustainability into industrial processes call for careful monitoring and evaluation systems [5, 6].

The industrial sector is changing greatly due to increased rivalry and declining availability of high-performance materials [7]. The requirement to switch to sustainable manufacturing techniques is great, as depending on a few energy resources aggravates technological and financial challenges [8]. Growing demand for long-lasting, ecologically friendly products forces material designers to pay attention to sustainability at every product development phase [9]. From conception to disposal, a sustainable design lessens the negative environmental impact of a product over its lifetime. This approach saves resources and meets customer demand for environmentally friendly products.

DOI: 10.1201/9781003633754-1

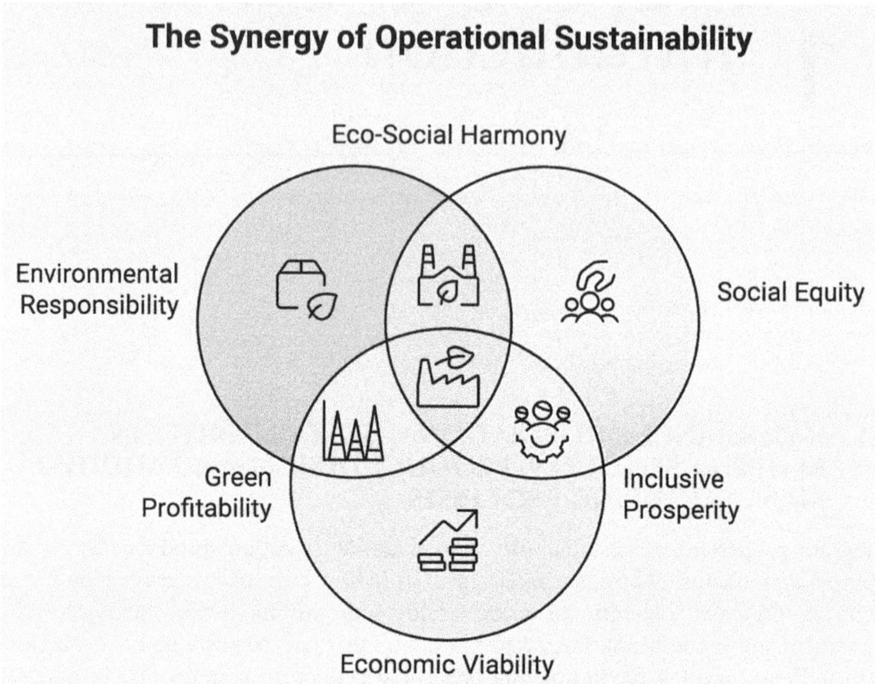

FIGURE 1.1 Triple Bottom Line approach to operational sustainability, highlighting the balance between environmental, economic, and social dimensions in manufacturing systems.

The interactions at all levels help one reach sustainability targets [10]. Sustainability is morally demanded and strategically crucial in the manufacturing industry, driving invention and innovation, and improving the resource economy. The integration of sustainability into production requires a thorough approach, including social, economic, and environmental aspects [11]. Such a thorough framework is essential to ensure that production techniques are efficient and socially and environmentally friendly. Although the development of industry and society depends on conventional production, sustainable needs are progressively driving change in it.

Furthermore, creativity and innovative technologies are critical for encouraging sustainable practices and are major drivers of the change to ecologically friendly manufacturing systems. Since it provides practical solutions to the ever-shifting needs of society and helps to adopt sustainable manufacturing techniques, research and development must thus be sustainably supported [12].

Lean and green technologies have helped the manufacturing sector move towards more ecologically friendly practices [13]. Whereas lean manufacturing focuses on waste reduction, green production aims to minimise its effects on the environment [14]. This integration allows manufacturers to boost earnings, gain a competitive edge, and satisfy social responsibility goals. Consumers seek long-lasting, ecologically friendly items since sustainability is becoming increasingly crucial in all goods and services.

From the earliest design to the final delivery, manufacturers must include sustainability issues in the whole product development process if they are to satisfy these needs and reduce the product's environmental impact over its whole life cycle [15]. This change means using strategies to extend the lifetime of products and including environmentally friendly materials and techniques [16]. Sustainable manufacturing techniques must be followed to lower the negative environmental impacts of industrial activities while maintaining economic competitiveness and production [17].

Modern sustainable manufacturing broadens the emphasis of manufacturing processes from cost and efficiency measures to encompass environmental stewardship and social responsibility, although historically, these areas have dominated concentration. Lowering carbon-intensive procedures helps manufacturers maximise their systems, develop more sustainable goods, and future-proof their company operations [4]. Given how carbon pricing systems are progressively influencing industrial supply chains, giving sustainability top priority in operations becomes not just moral but also strategically important. Using eco-design ideas can result in goods that satisfy needs and reduce environmental effects throughout their lifetimes [18]. Designed for durability, disassembly, and recyclability, eco-design techniques enhance product lifetime and lower waste.

Market trends demonstrating fast expansion in environmentally sensitive manufacturing technology highlight the necessity of sustainability even more. For example, the additive manufacturing sector grew by 18% in the past year, underscoring both the economic possibilities and the need to create environmentally friendly uses for this newly developed technology [19]. Due to this development path, manufacturers have obligations and opportunities to integrate sustainability concepts into their daily operations [20].

1.1.1 Operational Sustainability in Material Selection

Material selection is crucial for making manufacturing more sustainable, as it impacts the environment and resource utilisation. It entails prioritising materials with low environmental impact and ensuring their responsible sourcing and application. The selection of sustainable materials plays a pivotal role in decreasing environmental impacts and improving resource efficiency within manufacturing. Further, selecting sustainable materials is essential for reducing a product's environmental footprint, as materials significantly contribute to energy consumption and waste generation, both of which contribute to the environmental impact.

Materials should be sourced from suppliers who adhere to ethical and environmental impacts, ensuring that extraction, processing, and transportation practices minimise harm to ecosystems and communities. Materials obtained from renewable sources, such as bio-derived products used as raw materials, offer a viable alternative to conventional materials, reducing reliance on finite resources and minimising carbon emissions [18].

Using recycled materials reduces the need for raw material resources, decreases energy consumption during production, and diverts waste from landfills, contributing to a more circular economy [19]. Material choice is a key decision point where sustainability factors can greatly change the environmental impact of industrial activities. Choosing the appropriate material now requires balancing several factors, including environmental impact, durability, and contextual suitability, rather than focusing solely on structural efficiency [18, 20, 21].

Effective design concepts can drastically lower material demand, reducing environmental effects through careful material choice and specification [22, 23]. Materials should be chosen for their long-term durability and maintenance needs as well as for their immediate performance qualities. Products whose materials demand little maintenance and can be easily adapted help to greatly lower the environmental impact over the lifespan of the product.

By means of lifecycle analysis, environmental impact assessment helps manufacturers to examine the cradle-to-grave effect of construction materials, including embodied carbon, raw resource consumption, water consumption, and pollutant impacts. Beyond basic recycled content measures, this assessment offers a more realistic view of a material's actual sustainability credentials [3, 24]. Figure 1.2 illustrates the manufacturing lifecycle with sustainability integration.

The selection of sustainable materials depends, to a great extent, on local context factors. Particularly in distant places and developing nations where construction using foreign materials could generate long-term maintenance issues, materials

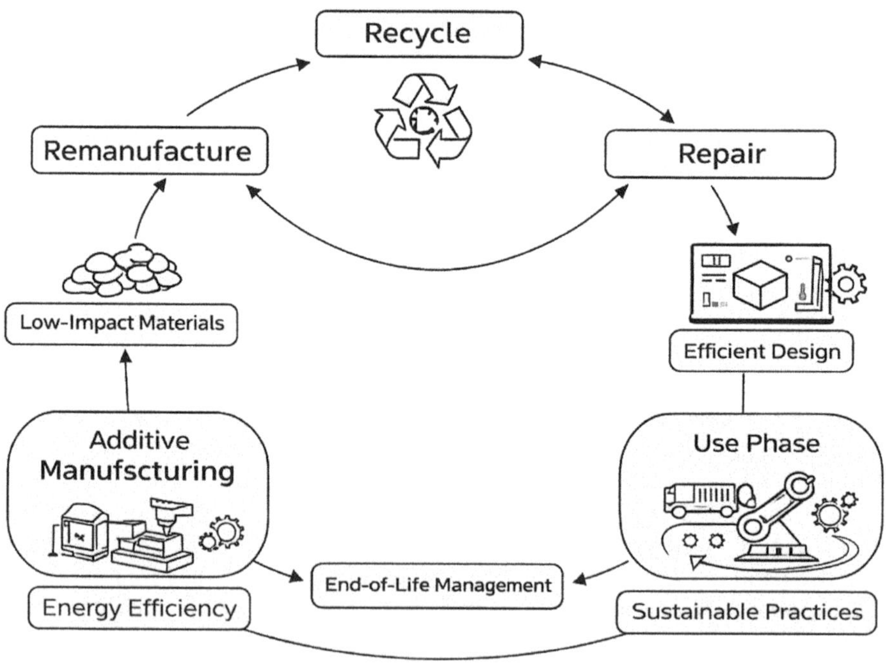

FIGURE 1.2 Conceptual framework of operational sustainability showing the interaction between material selection, product design, and manufacturing processes across the product life cycle.

should be acceptable for the climatic conditions and matched with the skills of the local workforce. This contextual approach guarantees that the advantages of sustainability are achieved over the lifetime of a product [25, 26].

Methodical decision-making techniques become vital when design teams must make ever more difficult choices about material choice. Many decisions taken in the real world are made without official procedures or careful thought, which frequently results in design waste and disputes. Structured methods like Choosing by Advantages (CVA) can help to solve this difficulty by fostering open and cooperative decision-making environments [3]. Design teams can methodically analyse and contrast many materials depending on pertinent criteria by using a structured decision-making framework, therefore guiding educated and sustainable material decisions [8]. Figure 1.3 illustrates the decision framework for sustainable material selection.

By combining several points of view, spotting pertinent sustainability factors, making clear trade-offs, recording decision rationales, separating value from cost, grouping location-dependent factors, and including supply chain considerations, the CVA method helps to select sustainable materials. This methodical approach guarantees that decisions match more general operational sustainability goals while helping design teams negotiate the complexity of sustainable material selection [27].

1.1.2 Operational Sustainability in Product Design

The design of a product defines a main intervention point for including sustainability across its lifetime [28, 29]. From conception to disposal, sustainable design approaches consider all aspects of a product's life cycle, generating possibilities to reduce environmental effects and maximise social and financial gains [3, 30].

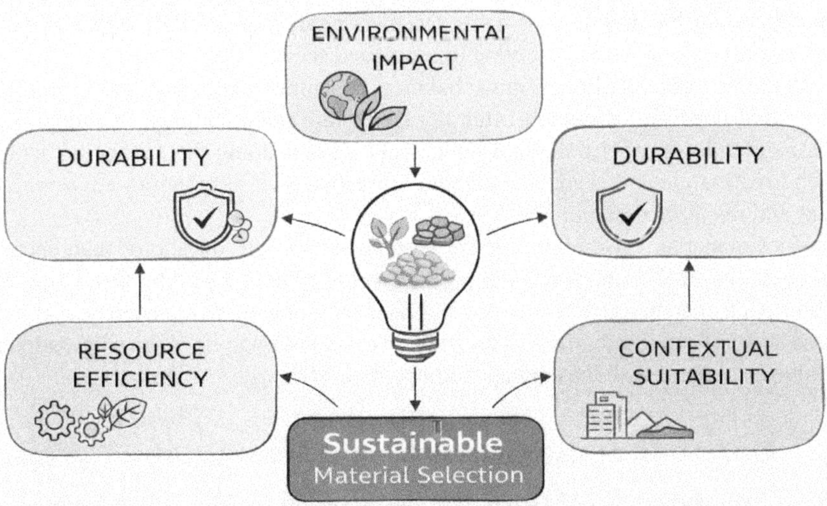

FIGURE 1.3 Decision framework for sustainable material selection.

Including sustainability in product design helps develop a holistic design and provides notable advancements. Important stages in this optimisation include environmental evaluation models and design improvement relying on feedback [24].

Choosing ecologically friendly materials, lowering energy use during manufacturing, designing products for a long lifetime, and promoting recyclability are among the eco-design concepts that boost sustainability. Early design choices impact both material and manufacturing choices and the whole product life cycle, including transportation, distribution, and end-of-life logistics, influencing sustainability [31].

Including sustainability in product creation will help businesses satisfy changing consumer demands, lower their environmental impact, and improve resource economy. Sustainable product development requires a multidisciplinary approach linking environmental experts, engineers, and designers. This cooperative environment facilitates the inclusion of sustainable ideas throughout the production of a product. Designers should also be able to access instruments and techniques for assessing areas of development possibilities and environmental impact. Industries can make wise judgements in line with their sustainability objectives using life cycle studies and eco-design technologies. Including sustainability concerns in product design demands a comprehensive and all-encompassing approach covering the complete product lifespan [9].

From design to disposal, sustainable product design covers a product's lifetime [30, 32]. This all-encompassing approach covers the use of limited resources in manufacturing, decreased waste and pollution, and guarantees that products may be safely disposed of at the end of their useful life. Early in the design phase, by adopting these ideas and producing goods that satisfy customer demand, businesses can greatly minimise their environmental impact [33].

The design process offers a special chance for proactive instead of reactive approaches to environmental solutions. Designers increasingly understand their obligation to actively contribute to minimise environmental damage through sensible design decisions. Applying sustainable design ideas helps designers produce goods that reduce waste and resource consumption while benefiting the earth and society [30, 33].

Sustainable product design must balance consumer expectations for improved features and capabilities, which often demand more manufacturing resources, with the demand to lower environmental impacts. Designers doing this balancing act must rely on creative ideas that meet pragmatic objectives while reducing environmental impact and resource demand [30, 33].

Besides material substitution, good sustainable design questions manufacturing techniques, end-of-life situations, and product architecture. Those who take these elements holistically can produce products that reduce environmental impact throughout their lifetime and perform impressively. This strategy turns sustainability from a limitation into a creative instrument with the engine of competitive advantage [24, 33].

1.1.3 Operational Sustainability in Additive Manufacturing Processes

Additive manufacturing (AM) has become a strong enabler of operational sustainability with specific advantages over traditional production techniques. The layer-by-layer building method inherent in AM presents opportunities for material savings, design optimisation, and supply chain change that can significantly increase sustainability outcomes [34].

Reduced buy-to-fly ratios and tooling needs help to improve material efficiency, one of AM's most significant environmental advantages. While conventional subtractive manufacturing techniques sometimes lose significant volumes of material, AM techniques use just the material needed to generate the end product, therefore considerably reducing waste streams and raw material usage [31].

Using AM technologies produces several levels of energy economy. Energy requirements are lowered in manufacturing by lessening the material supply and design optimisation. Moreover, AM-enabled design innovations that allow more efficient energy or fuel consumption throughout the use phase of a product help the operational stage, therefore extending sustainability benefits outside the production line itself [5].

Production distribution allows AM technologies to maximise supply chains, contributing even more to sustainability. On-site manufacturing capabilities help to reduce transportation-related emissions and energy use by limiting the demand for long-distance products and material transportation. This distributed approach reduces supply chain risks and increases resilience, even raising sustainability performance [4, 6, 35].

One often overlooked advantage of AM technology is product life extension via highly tuned or customised components. Building components tailored for defined uses lets AM enable longer-lasting, better-performing products needing less frequent replacement. Reducing the manufacturing replacement product resource intensity helps to increase the sustainability benefits of AM [36, 37].

Using a circular economy approach will help to maximise the opportunities for the sustainability of AM technology. This approach balances social, financial, and environmental challenges by stressing waste reduction through continual circulation of resources. AM technologies are notably fit to support circular economy concepts by means of material reuse, design for disassembly, and on-demand manufacturing that lowers overproduction waste [27].

The circular economy model provides a whole prism through which one may evaluate and enhance the sustainability performance of AM operations. Designing goods for circularity from the start and using AM's special qualities will help manufacturers build closed-loop systems that reduce waste generation and increase resource use throughout the product life [2, 27, 33, 37].

Though AM has a lot of promise, following sustainability activities needs constant assessment and improvement. It is important to completely capture the environmental effects of AM, especially when assessing its energy usage, material selection, and waste management procedures, by doing lifecycle assessments. Figure 1.4 shows Life cycle assessment (LCA) stages for materials.

Furthermore, working together and exchanging best practices within the AM community, it is possible to speed up the shift to more sustainable procedures and maximise the technology's potential for good environmental impact. Figure 1.5 illustrates an overview of an additive manufacturing process.

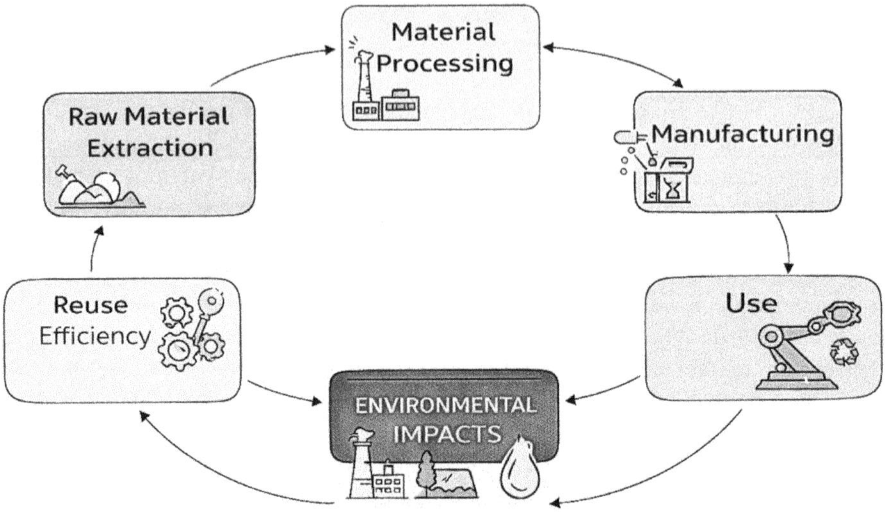

FIGURE 1.4 Life cycle assessment (LCA) stages for materials.

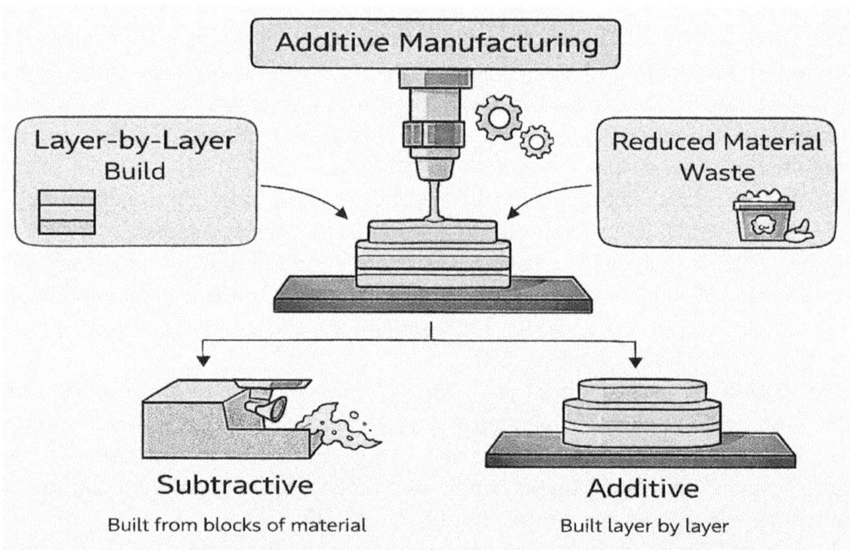

FIGURE 1.5 Additive manufacturing process overview.

1.2 ENVIRONMENTAL-RELATED INTERNATIONAL REGULATIONS AND SUSTAINABILITY

Environmental issues spanning national lines and requiring coordinated global responses include pollution, biodiversity loss, and climate change [8, 38]. Under this paradigm, formal treaties and agreements among nations, international environmental rules, become essential instruments for addressing shared environmental issues and advancing sustainability. These accords build a foundation for worldwide environmental governance by means of encouragement of cooperation, setting standards, and national environmental policy guidance [5].

Agreements for global environmental preservation are supposed to: (i) make legally or politically enforceable pledges for environmental preservation, (ii) organise worldwide projects aiming at lowering transboundary environmental dangers, (iii) promote sustainable development by combining financial and social objectives with ecological integrity, and (iv) provide ways of monitoring, compliance, and information sharing. By means of international cooperation, these guidelines enhance the capacity of the world society to address interrelated and demanding environmental problems [25].

1.2.1 Key Worldwide Environmental Agreements

The United Nations Framework Convention on Climate Change (UNFCCC), the 1992 Convention on Climate Change United Nations Framework, is an international treaty meant to stabilise atmospheric greenhouse gas concentrations in order to reduce negative anthropogenic intervention with the climate system. It paved the path for other accords, including the following:

Under the UNFCCC, the legally enforceable Paris Agreement (2015) pledges countries to limit the worldwide temperature rise to well below 2°C with policies aiming at ceilings at 1.5°C. It emphasises transparency, nationally defined contributions (NDCs), and creation of adaptation capacity.

Kyoto Protocol (1997): Designed as the first major UNFCCC addition, the Kyoto Protocol set legally enforceable targets for rich countries to reduce greenhouse gas emissions. Though on a limited scale, it was a major first towards quantifiable emissions reductions.

Quebec Protocol (1987): This pact deals with the phase-down of substances causing ozone depletion (ODS). Considered one of the most effective environmental agreements, it greatly helps the ozone layer rebuild and avoids negative effects on health, including skin cancer and cataracts.

Basel Convention (1989): The Basel Convention seeks to regulate hazardous waste transboundary movement and disposal. Particularly to safeguard sensitive areas from harmful waste disposal, it aims to reduce waste generation and guarantee environmentally friendly management of hazardous chemicals.

Stockholm Convention (2001): This convention aims to eradicate or limit the manufacturing and application of persistent organic pollutants (POPs). Because of their resistance to degradation and bioaccumulative characteristics, POPS present long-term hazards to human health and the environment.

Rotterdam Convention (1998): Emphasising previous informed consent, this agreement mandates exporters notify and get permission from importing nations before engaging in trade of specific hazardous chemicals, therefore encouraging informed decision-making and safety.

Convention on Biological Diversity (CBD, 1992): The CBD promotes the conservation of biological diversity, sustainable use of its components, and fair and equitable sharing of benefits arising from genetic resources. It is crucial in halting biodiversity loss and integrating conservation into development planning.

Convention on International Trade in Endangered Species (CITES, 1973): CITES controls the trading of threatened species internationally to guarantee their existence. It offers a legislative basis to stop illegal wildlife trafficking and overuse.

These multinational models are important for world sustainability since they set Norms and Standards, offering guidance for developing national laws and policies. Encouragement of Multilateral Cooperation: Understanding environmental problems worldwide and calling for group answers. Encouraging Sustainable Development: Integrating environmental protection with social and economic dimensions. By means of reporting responsibilities, reviewing systems, and holding conferences with the parties (COPs), one can improve accountability and monitoring.

Despite their importance, several challenges hinder the effective implementation of these treaties: Compliance and Enforcement: National-level enforcement mechanisms vary widely, affecting treaty efficacy. Resource Constraints: Developing countries often lack the financial and technical capacity to fulfil obligations. Evolving Environmental Threats: Emerging issues such as microplastic pollution, environmental displacement, and digital environmental monitoring necessitate continual treaty revision and innovation. Political Commitment: Shifts in political priorities and international relations can impact the continuity and strength of environmental governance.

International environmental regulations are indispensable to achieving global sustainability goals. They serve as instruments of legal obligation and platforms for knowledge exchange, cooperation, and innovation. Strengthening these frameworks, ensuring equitable participation, and adapting them to evolving scientific knowledge and socio-economic realities remain crucial. As the world grapples with escalating environmental crises, the role of such international agreements becomes ever more critical in securing a resilient and sustainable future for all [3, 5].

REFERENCES

1. V. T. Dao, I. M. Langella, & J. A. Carbo, From green to sustainability: Information Technology and an integrated sustainability framework. *The Journal of Strategic Information Systems*, 20(1) (2011) 63.
2. D. L. de M. Nascimento, V. Alencastro, O. L. G. Quelhas, R. G. G. Caiado, J. A. Garza-Reyes, L. Rocha-Lona, et al., Exploring industry 4.0 technologies to enable circular economy practices in a manufacturing context. *Journal of Manufacturing Technology Management*, 30(3) (2018) 607.
3. P. Saxena, P. Stavropoulos, J. D. Kechagias, & K. Salonitis, Sustainability assessment for manufacturing operations. *Energies*, 13(11) (2020) 2730.
4. M. A. Rosen, & H. A. Kishawy, Sustainable manufacturing and design: Concepts, practices and needs. *Sustainability*, 4(2) (2012) 154.
5. P. Nyamekye, R. Lakshmanan, V. Tepponen, & S. Westman, Sustainability aspects of additive manufacturing: Leveraging resource efficiency via product design optimization and laser powder bed fusion. *Heliyon*, 10(1) (2023) e23152 (1–21).

6. C. G. Mançanares, E. Zancul, & P. A. C. Miguel, Sustainable manufacturing strategies: A literature review on additive manufacturing approach. *Product Management & Development*, 13(1) (2015) 47.
7. S. Rouf, A. Malik, N. Singh, A. Raina, N. Naveed, M. I. H. Siddiqui, et al., Additive manufacturing technologies: Industrial and medical applications. *Sustainable Operations and Computers*, 3 (2022) 258–274.
8. M. Gopal, H. G.Lemu, & E. M. Gutema, Sustainable additive manufacturing and environmental implications: Literature review. *Sustainability*, 15(1) (2022) 504.
9. A. Santanaa, P. Afonsoa, A. Zaninb, & R. Wernkeb, Costing models for capacity optimization in Industry 4.0: Trade-off between used capacity and operational efficiency. *Procedia Manufacturing*, 13 (2017) 1183–1190.
10. H. A. Kishawy, H. Hegab, & E. M. A. Saad, Design for sustainable manufacturing: Approach. *Implementation, and Assessment. Sustainability*, 10(10) (2018) 3604.
11. R. R. Delgadillo, K. Medini, & T. Wuest, A DMAIC framework to improve quality and sustainability in additive manufacturing—A case study. *Sustainability*, 14(1) (2022) 581.
12. M. Garetti, & M. Taisch, Sustainable manufacturing: Trends and research challenges. *Production Planning & Control*, 23 (2011) 83–104.
13. A. Sharma, Sustainable manufacturing–the past, present and future. *International Journal of Engineering Applied Sciences and Technology*, 6(2) (2021) 62–66.
14. R. K. Lukman, V. Omahne, & D. Krajnc, Sustainability assessment with integrated circular economy principles: A toy case study. *Sustainability*, 13(7) (2021) 3856.
15. M. Jin, R. Tang, Y. Ji, F. Liu, L. Gao, & D. Huisingh, Impact of advanced manufacturing on sustainability: An overview of the special volume on advanced manufacturing for sustainability and low fossil carbon emissions. *Journal of Cleaner Production*, 161 (2017) 69–74.
16. V. Kumar, & K. Shahin, Artificial intelligence and machine learning for sustainable manufacturing: Current trends and future prospects. *Intelligent and Sustainable Manufacturing*, 2(1) (2025) 10002.
17. P. Chen, I. L. Blanco, J. F. Santos, & P. R. de A. López, Lean manufacturing and environmental sustainability: The effects of employee involvement, stakeholder pressure and ISO 14001. *Sustainability*, 12(18) (2020) 7258.
18. C. Vezzoli, *The "Material" Side of Design for Sustainability*. Elsevier eBooks, (Elsevier BV, 2013), p. 105.
19. S. Jung, L. B. Kara, Z. Nie, T. W. Simpson, & K. S. Whitefoot, Is additive manufacturing an environmentally and economically preferred alternative for mass production? *Environmental Science & Technology*, 57(16) (2023) 6373.
20. A. Giallanza, Industry 4.0 technologies for manufacturing sustainability: A systematic review and future research directions. *Applied Sciences*, 11(12) (2021) 5725.
21. A. Simboli, R. Taddeo, & A. Morgante, Value and wastes in manufacturing. An overview and a new perspective based on eco-efficiency. *Administrative Sciences*, 4(3) (2014) 173.
22. C. Collin, G. Olesen, & A. Q. Secher, A case-based study on the use of life cycle assessment and life cycle costing in the building industry. In *IOP Conference Series Earth and Environmental Science* (IOP Publishing, 2019), p. 12163.
23. S. Barbhuiya, & B. B. Das, Life cycle assessment of construction materials: Methodologies, applications and future directions for sustainable decision-making. *Case Studies in Construction Materials*, 19 (2023) e02326.
24. Y. Tang, K. Mak, & Y. F. Zhao, A framework to reduce product environmental impact through design optimization for additive manufacturing. *Journal of Cleaner Production*, 137 (2016) 1560.

25. I. Ribeiro, F. Matos, C. Jacinto, H. Salman, G. Cardeal, H. Carvalho, et al., Article framework for life cycle sustainability assessment of additive manufacturing. *Sustainability*, 12(3) (2020) 929.
26. F. Fruggiero, A. Lambiase, R. Bonito, & M. Fera, The load of sustainability for additive manufacturing processes. *Procedia Manufacturing*, 41 (2019) 375.
27. G. Taddese, S. Durieux, & E. Duc, Sustainability performance indicators for additive manufacturing: A literature review based on product life cycle studies. *The International Journal of Advanced Manufacturing Technology*, 107 (2020) 3109.
28. H. Tian, H. Zhang, & T. Liu, Research on material selection of product design under environmental awareness. In *IOP Conference Series Materials Science and Engineering* (IOP Publishing, 2019), p. 12057.
29. A. Y. Soliman, & A. M. Al-Bahi, A win-win strategy to integrate sustainability objectives in product design – An educational approach. *International Journal of Engineering Pedagogy (iJEP)*, 8(5) (2018) 29.
30. L. Y. Ljungberg, Materials selection and design for development of sustainable products. *Materials & Design (1980–2015)*, 28(2) (2005) 466.
31. K. Ramani, D. Ramanujan, W. Z. Bernstein, F. Zhao, J. W. Sutherland, C. A. Handwerker, et al., Integrated sustainable life cycle design: A review. *Journal of Mechanical Design*, 132(9) (2010) 1–5.
32. T. S. Singhal, J. K. Jain, D. A. Ramacharyulu, A. P. Jain, D. S. Abdul-Zahra, M. Manjunatha, et al., Eco-design of products and processes: A review on principles and tools for sustainable manufacturing. *E3S Web of Conferences*, 505 (2024) 1033.
33. H. Hegab, N. Khanna, N. Monib, & A. Salem, Design for sustainable additive manufacturing: A review. *Sustainable Materials and Technologies*, 35 (2023) e00576.
34. T. Peng, K. Kellens, R. Tang, C. Chen, & G. Chen, Sustainability of additive manufacturing: An overview on its energy demand and environmental impact. *Additive Manufacturing*, 21 (2018) 694.
35. R. Godina, I. Ribeiro, F. Matos, B. T. Ferreira, H. Carvalho, & P. Peças, Article impact assessment of additive manufacturing on sustainable business models in industry 4.0 context. *Sustainability*, 12(17) (2020) 7066.
36. M. Despeisse, & S. Ford, The role of additive manufacturing in improving resource efficiency and sustainability. In IFIP International Conference on Advances in Production Management Systems (APMS), September 2015, Tokyo, Japan (Springer Science+ Business Media, 2015), pp. 129–136.
37. S. Kondoh, T. Tateno, Y. Kishita, H. Komoto, & S. Fukushige, The potential of additive manufacturing technology for realizing a sustainable society. In Matsumoto, M., Masui, K., Fukushige, S., & Kondoh, S. (eds), *Ecoproduction* (Springer International Publishing, 2016), p. 475.
38. M. K. Niaki, S. A. Torabi, & F. Nonino, Why manufacturers adopt additive manufacturing technologies: The role of sustainability. *Journal of Cleaner Production*, 222 (2019) 381.

2 Material Selection for Operational Sustainability in Additive Manufacturing

This chapter explores the significance of material selection in achieving operational sustainability in additive manufacturing. It investigates the integration of low-energy, recyclable, biodegradable, and bio-based materials with sustainable practices. The chapter highlights the use of life cycle assessment as a fundamental methodology for assessing the environmental effects of a product from the extraction of raw materials to its end of life. Additionally, it discusses material efficiency techniques that minimise waste and resource consumption without sacrificing functionality, such as lightweighting, topology optimisation, and generative design. It also assesses the performance, compatibility, and recyclability of metal powders and eco-friendly filaments, emphasising their significance. The subject is broadened to cover process optimisation, additive integration, and the application of automation and digitisation to decrease the environmental impacts. The chapter adopts a comprehensive approach to sustainable material selection, establishing it as a key component in the development of future-ready additive manufacturing systems.

2.1 INTRODUCTION TO SUSTAINABLE MATERIAL SELECTION

The selection of materials plays a crucial role in determining the operational sustainability of products developed through additive manufacturing [1, 2]. This selection process requires a comprehensive understanding of material properties, manufacturing processes, and the environmental impact throughout the product's life cycle [3]. Life cycle assessment systematically evaluates the environmental burdens associated with a material or product from cradle to grave, including raw material extraction, manufacturing, transportation, use, and end-of-life disposal [4]. Designers and engineers can use life cycle assessment to identify opportunities to minimise environmental impacts at each stage of a product's life cycle [5].

To make informed material choices, it is essential to evaluate multiple factors, including environmental, economic, and functional considerations. A key tool in this evaluation is the life cycle assessment (LCA), which provides a comprehensive analysis of a material's environmental impact. LCA considers every stage of the material's life, beginning with raw material extraction, then processing, manufacturing, use, and finally end-of-life disposal or recycling. The process of responsible material selection

DOI: 10.1201/9781003633754-2

begins by addressing three critical questions: Is the material derived from recycled or bio-based sources, and what is the energy footprint associated with its extraction and processing? These questions help determine whether the material supports low-impact, sustainable sourcing [6, 7].

Can the material be reused, recycled, or composted? Does it degrade safely, or does it release harmful substances during breakdown? Understanding a material's post-use implications is vital for circular economy practices. This encompasses the material's carbon emissions, water consumption, pollution potential, and overall impact on ecosystems and biodiversity. Tools like LCA are essential for quantifying these variables and making evidence-based decisions. The integration of sustainable materials is paramount for enhancing operational sustainability, and this involves prioritising materials with a minimal environmental footprint throughout their life cycles [8]. Sustainable materials often possess characteristics such as being renewable, recyclable, biodegradable, or sourced from sustainably managed resources.

Optimising material usage is another critical aspect of material selection for additive manufacturing, directly impacting waste reduction and resource utilisation [9]. By employing design strategies such as lightweighting, topology optimisation, and generative design, engineers can create parts that use only the necessary amount of material while maintaining structural integrity and performance requirements [10].

Additive manufacturing processes offer unique opportunities for material optimisation, allowing for the creation of complex geometries and customised parts that minimise waste and maximise material efficiency. Furthermore, selecting appropriate filaments and powders is crucial for additive manufacturing processes [11]. The relationship between additive manufacturing and the circular economy is complex and not yet fully understood [12]. While additive manufacturing offers the potential for localised production and reduced material consumption, it is not clear whether current practices promote truly circular material flows [12].

Additive manufacturing is suited to support multiple product life cycles and can provide valuable contributions to a circular economy [13]. The use of sustainable materials, those that are recyclable, biodegradable, or have a reduced carbon footprint, is increasingly vital. These materials help minimise environmental harm, lower resource consumption, and support long-term sustainability targets.

Optimising material usage is equally important. Product designs that reduce material consumption without compromising functionality or durability can greatly enhance resource efficiency. Additive manufacturing enables such optimisation through design flexibility and precision. Material choices, such as filaments and powders in 3D printing, must meet two essential criteria: compatibility with the manufacturing process and environmental responsibility. Selecting materials that meet both requirements ensures effective production and reduces negative environmental outcomes.

Furthermore, advanced manufacturing techniques, such as 3D printing and automated fabrication, can significantly reduce material waste. These technologies are most effective when paired with sustainable materials tailored to their capabilities. A holistic approach to material selection is necessary, considering environmental, economic, and social impacts alongside technical performance. As sustainability becomes a central goal, key drivers are pushing this shift: Climate change necessitates reducing the carbon footprint of materials and manufacturing processes. Pollution requires careful material choices to avoid harmful emissions into the air, water, and soil. Resource depletion makes the efficient use of materials, particularly recycled or renewable ones, essential to preserve finite natural resources. Adopting strategies that ensure sustainability and minimise environmental impact is essential, with smart manufacturing systems playing a key role in achieving these goals [14].

The need for substantial changes based on environmental concerns forces the manufacturing industry to face economic and technological hurdles due to the use of finite materials and energy resources [11].

2.2 LIFE CYCLE ASSESSMENT

Life cycle assessment (LCA) is a vital tool for evaluating the environmental impacts of materials, processes, or products across their entire life span, from raw material extraction to end-of-life management [11]. It provides a structured and scientific basis for identifying environmental hotspots, guiding material selection, and supporting sustainable decision-making in product design and manufacturing. This 'cradle-to-grave' methodology helps identify opportunities to reduce environmental burdens at every stage of a product's life cycle, making it especially relevant for additive manufacturing [10, 15]. Figure 2.1 illustrates the environmental impacts of a material or product throughout its entire lifecycle.

LCA typically examines five major phases of the life cycle. The first is raw material extraction and processing, which includes sourcing materials from natural resources, transporting them, and converting them into usable forms through often energy-intensive methods [16]. The next stage is manufacturing, where raw inputs are transformed into finished products. In the context of additive manufacturing, this covers filament or powder production, 3D printing operations, and post-processing treatments such as cleaning, support removal, and heat treatment. Following production is distribution and transportation, which encompasses packaging, storage, and delivery to the end user [16].

The fourth phase is use and maintenance, which includes the energy, resources, and effort required to operate and maintain the product during its service life. Finally, the end-of-life stage evaluates what happens to the product once it is no longer in use, whether it is reused, recycled, incinerated, or sent to a landfill. Assessing each of these stages allows designers and manufacturers to pinpoint where most environmental impact occurs and explore ways to reduce it [6, 17].

LCA uses a range of environmental indicators to provide a holistic picture of impact. These indicators may include carbon footprint, water consumption, energy use, air and water emissions, and resource depletion. By analysing these metrics, LCA supports more sustainable decisions during material selection, helping manufacturers opt for materials and processes with lower environmental costs [18].

The methodology itself consists of four core steps: Goal and Scope Definition – establishing the purpose of the assessment, system boundaries, and the functional unit used for comparison. Inventory Analysis – collecting data on all material and energy

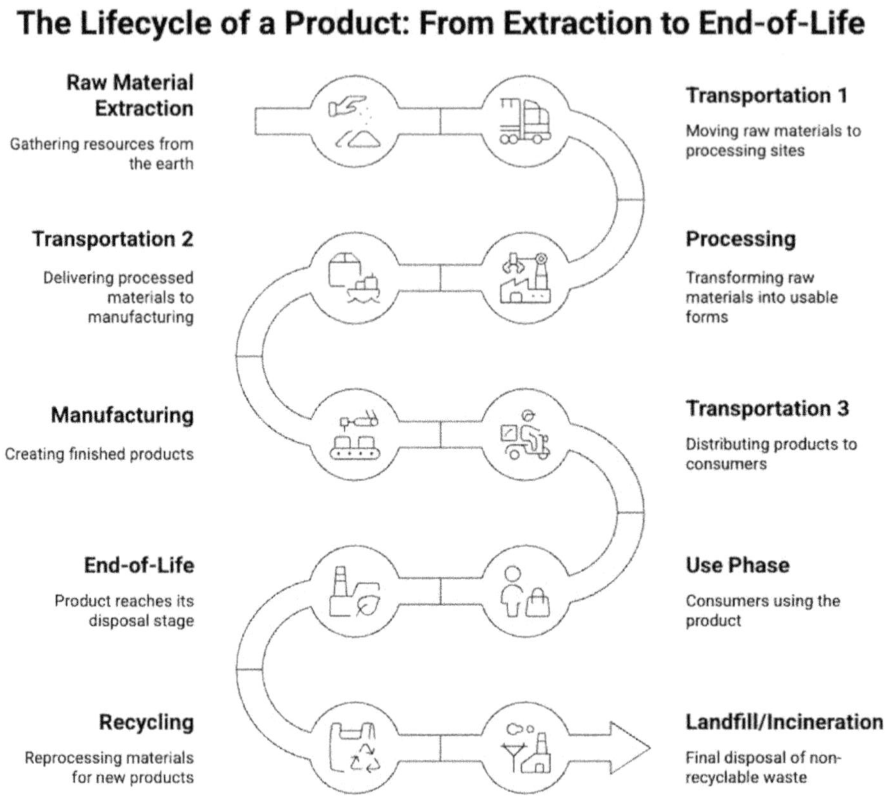

FIGURE 2.1 A visual representation of the environmental impacts of a material or product throughout its entire lifecycle.

inputs and environmental outputs at each life cycle stage. Impact Assessment – analysing how the inputs and outputs affect various ecological categories, including climate change, acidification, and resource depletion. Interpretation – concluding the results, identifying key hotspots, and proposing actions for improvement.

In additive manufacturing, LCA plays a crucial role. It helps identify which phases of the process contribute most to environmental harm, whether it is energy used during printing, emissions from material production, or waste from post-processing. By doing so, manufacturers can take targeted steps to improve sustainability [2, 11].

LCA is also used to compare material options for similar functions. For example, it can help determine whether titanium or aluminium powder has a smaller environmental impact in an aerospace application. Similarly, LCA can evaluate how different print parameters, such as laser power, layer thickness, and build orientation, affect energy use and emissions. These insights inform both material selection and process optimisation, ultimately enabling more sustainable production strategies [11, 19].

Another important application is in design for sustainability. LCA can be used to assess how design changes, such as altering infill structures or minimising supports, can reduce material use, improve recyclability, or extend product life. It also allows comparison across different end-of-life scenarios, such as reuse, recycling, and landfilling, to support circular economy models in additive manufacturing [10].

Beyond the factory floor, LCA serves a broader purpose. It supports environmentally conscious design by enabling engineers to quickly assess different design and production options, even as the methodology itself continues to evolve and become more user-friendly. In sectors like construction, LCA is also being used to guide the use of sustainable materials and building practices, helping reduce environmental impacts on a much larger scale [11].

Importantly, the integration of LCA into engineering education is essential for building long-term sustainability awareness. Teaching LCA principles helps future professionals develop critical thinking skills and a deeper understanding of resource management, fostering a mindset of environmental responsibility [6, 20].

In summary, life cycle assessment is not just a technical tool but a foundational element in sustainable engineering. In additive manufacturing, it supports better material choices, smarter design, and cleaner production. Its ability to offer a complete view of environmental performance makes it indispensable for driving operational sustainability and shaping future practices.

2.3 SUSTAINABLE MATERIALS

Sustainable materials play a critical role in advancing operational sustainability in additive manufacturing by reducing environmental impact and conserving resources. These materials are typically recyclable, biodegradable, or derived from renewable sources, offering viable alternatives to conventional materials that contribute to pollution and resource depletion. By incorporating sustainable materials into both product design and manufacturing processes, organisations can lower their carbon footprint, conserve energy, and reduce overall waste [6].

The adoption of sustainable materials supports the principles of the circular economy by closing material loops, reducing reliance on virgin resources and minimising end-of-life burdens. Beyond environmental benefits, the use of such materials contributes to a company's social sustainability profile, particularly when paired with responsible sourcing and safe production practices. Social sustainability assessments increasingly consider factors such as worker health and safety, energy and water use, and emissions at industrial sites [11, 21].

In additive manufacturing (AM), sustainable materials are those that reduce environmental impact throughout their entire life cycle, from raw material extraction through processing, use, and disposal. These materials are selected not only for their technical performance but also for their life cycle efficiency, which includes minimising waste and energy consumption. Importantly, the choice of material directly influences the sustainability of the entire manufacturing operation, making it a central design consideration [19, 22]. Figure 2.2 illustrates the flow of materials through the additive manufacturing process.

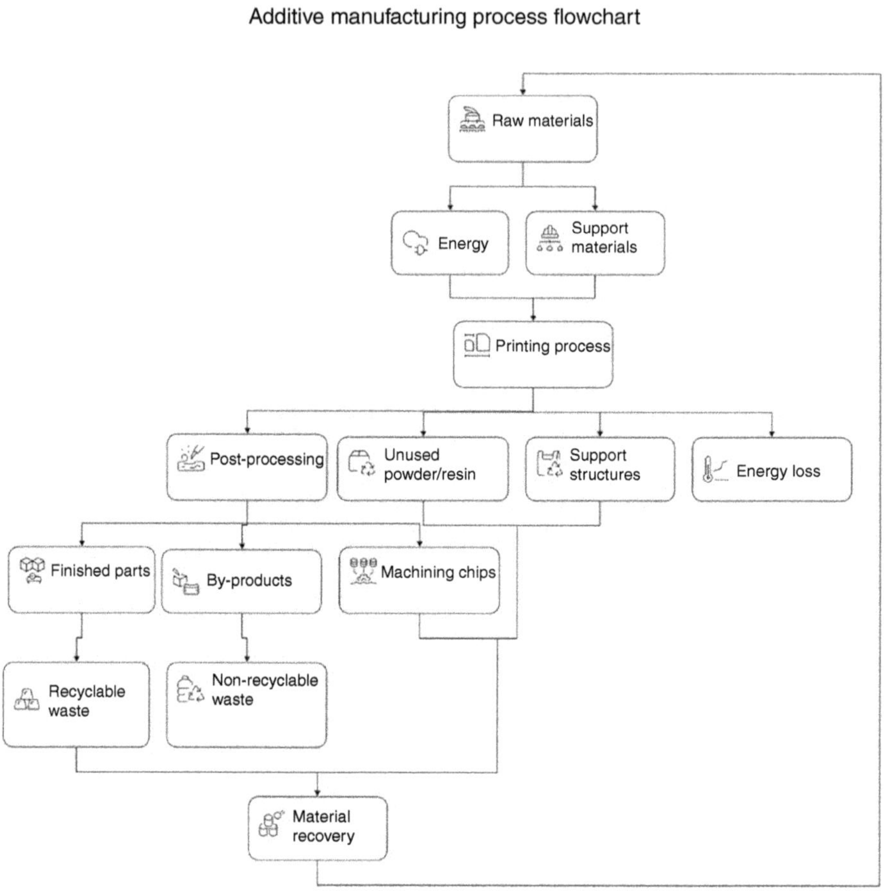

FIGURE 2.2 A diagram showing the flow of materials through the additive manufacturing process.

A variety of material classes have emerged as sustainable options for AM. Each category offers specific environmental and technical advantages but also comes with important trade-offs and limitations [16, 23]. The following are the categories of sustainable materials.

1. *Recyclable Materials:* Recyclable materials can be reprocessed and reused in the production of new parts, helping reduce the demand for virgin feedstocks. Common recyclable thermoplastics in AM include acrylonitrile butadiene styrene (ABS), polylactic acid (PLA), polypropylene (PP), and polyethene terephthalate glycol-modified (PETG). In metal additive manufacturing, aluminium, titanium, and stainless-steel powders can also be reused, though metal recycling often requires significant energy. The actual recyclability of a material depends on local

infrastructure, the purity of recovered material, and contamination levels, which may degrade material quality [10, 16, 24].

2. *Biodegradable Materials:* Biodegradable materials are capable of decomposing into natural substances such as carbon dioxide, water, and biomass through microbial activity. PLA is a notable example, widely used in desktop 3D printing for its biodegradability and ease of processing. However, biodegradability is highly dependent on environmental conditions such as heat, moisture, and microbial presence. Most biodegradable plastics require industrial composting facilities to degrade effectively, limiting their end-of-life options in typical disposal systems [25].
3. *Bio-Based Materials:* Bio-based materials are made from renewable biological sources such as plant-based oils, starches, or cellulose. Examples include bio-based polyamides, polyurethanes, and epoxies. These materials can reduce dependence on fossil fuels and are generally associated with lower carbon emissions. However, their sustainability also depends on upstream agricultural impacts, including land use, irrigation, and chemical inputs such as fertilisers and pesticides [26].
4. *Low-Energy Materials:* Low-energy materials are those that require less energy for processing compared to traditional options. Polymers, for instance, generally consume less energy during production and forming than metals. Materials that can be processed at lower temperatures or require fewer manufacturing steps fall into this category. Nonetheless, energy use should be assessed across the full life cycle, from raw material extraction to product disposal, to ensure meaningful reductions in environmental impact [10, 23]. Figure 2.3 illustrates energy consumption across major additive manufacturing processes, based on typical industry data (kWh per kg of material processed).

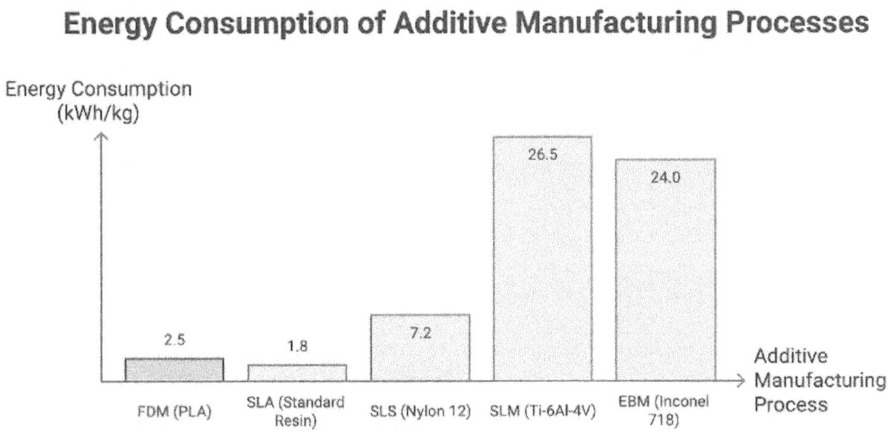

FIGURE 2.3 A representation of energy consumption across major additive manufacturing processes, based on typical industry data (kWh per kg of material processed).

In addition to the core material types listed above, current research and industrial practice are exploring several promising alternatives and emerging trends to enhance sustainability in additive manufacturing [23].

Renewable Polymers: These include materials derived from sources such as cellulose, corn starch, and vegetable oils. They provide a renewable alternative to petrochemical-based plastics while maintaining functional properties [27].

Alternative Metals: Lightweight metals such as aluminium and magnesium are being explored as substitutes for more energy-intensive options like steel and titanium. Their lower processing energy requirements make them attractive for sustainable AM [23].

Composites: Sustainable composite materials that combine recycled or bio-based fillers with high-performance polymer matrices are being developed to balance mechanical performance with environmental responsibility [28].

Additives: Additives play a critical role in enhancing material properties, but some, such as phthalates, pose environmental and health risks. The shift towards bio-based and biodegradable additives presents a path towards more eco-friendly formulations [29].

Figure 2.4 illustrates emerging sustainable materials with advantages and limitations.

Emerging Sustainable Alternatives

Material	Key Advantage	Current Limitation
Wood-PLA Composite	30% lower CO2 than PLA	50% weaker than pure PLA
Algae-Based Resin	Carbon-negative production	Limited color options
Recycled Nylon	45% energy savings vs virgin nylon	15% strength reduction
PHA Biopolymer	Marine-degradable in 6 months	2x cost of PLA

FIGURE 2.4 An emerging sustainable alternative material with advantages and limitations.

2.3.1 Challenges and Considerations

While sustainable materials offer numerous advantages, they often present trade-offs between durability and degradability. Materials must maintain structural integrity over a product's useful life but also degrade responsibly at the end-of-life stage. Achieving this balance remains a technical challenge and an active area of research [30].

Moreover, choosing sustainable materials does not automatically result in sustainable products. The full environmental benefit depends on how the materials are sourced, processed, and disposed of, as well as their compatibility with AM processes. In this context, LCA becomes a valuable tool for evaluating materials not only from a performance and cost standpoint but also from an environmental perspective [10].

By carefully considering the type, source, and life cycle performance of materials, manufacturers can make more informed choices that align with sustainability goals. This contributes to more responsible use of resources and supports long-term strategies for circular production and low-impact manufacturing.

2.4 MATERIAL EFFICIENCY AND DESIGN STRATEGIES

Material efficiency in additive manufacturing is achieved through design optimisation and waste reduction [23]. It plays an important role in sustainable manufacturing. By improving material efficiency, manufacturers can reduce waste, save energy, and lower their environmental impact [31]. It means using only the amount of material needed to make a strong, durable product, without reducing its quality or function. This approach helps both the environment and production costs. Key methods that support material efficiency include lightweight design, topology optimisation, and generative design [32].

In traditional manufacturing, materials are often used inefficiently. The 'take-make-dispose' model creates waste and uses more energy than needed. Additive manufacturing offers a better approach. It builds parts layer by layer, using only the required amount of material. This reduces waste at the source. But to fully benefit from this, designs must be made with efficiency in mind. When material efficiency is included early in the design phase, it leads to better results throughout the product's life, from production to end of life [16].

Additive manufacturing supports material efficiency by its nature. Unlike subtractive methods, it doesn't require cutting away large sections of raw material. Instead, it creates parts with precise shapes and minimal excess. This allows engineers to produce designs that would be difficult or impossible to make using traditional methods. To improve this further, three main design strategies are used: lightweighting, topology optimisation, and generative design [21].

Lightweighting involves reducing the weight of a part while keeping its strength and function. This is especially important in industries like aerospace and automotive, where reducing weight helps cut fuel use and emissions [33]. Lightweighting can be done by choosing strong, low-density materials, designing thin-walled or hollow parts, and adding internal lattice structures. These approaches reduce the total amount of material used, making parts lighter and more efficient [8]. Additive manufacturing supports lightweighting because it can produce complex, hollow, or internal structures that are difficult to achieve with traditional tools [10].

Topology optimisation is another method that supports material efficiency. It uses software to calculate the best way to distribute material within a part. The software removes unnecessary material from areas that don't carry much load, resulting in lighter and stronger parts [33]. These optimised designs often look irregular or organic and are difficult to produce using standard manufacturing methods. With additive manufacturing, these complex shapes can be produced accurately, allowing engineers to use material only where it is needed most [34].

Generative design is a computer-based approach that creates multiple design options based on a set of goals and limits. The designer inputs information such as material type, load requirements, and manufacturing method. The software then explores many possible shapes that meet those needs [35]. Often, these designs are too complex for traditional methods but are ideal for additive manufacturing. This combination helps improve material use and creates new design possibilities [36].

Using these design strategies helps improve sustainability in several ways. Reducing material use means fewer raw materials need to be extracted and processed. Lighter parts also use less energy during transport and use. Less waste is produced during manufacturing, and parts can often last longer because they are designed more effectively. When parts are designed for disassembly or reuse, it's easier to recycle materials at the end of a product's life. These steps all contribute to better resource use and lower environmental impact [21].

Several industries are already using these strategies. In aerospace, structural parts and brackets have been redesigned using topology optimisation and additive manufacturing. This has led to weight reductions of up to 50% while still meeting safety standards. In the automotive sector, generative design is used to create lighter suspension systems and brackets that improve fuel efficiency. In medicine, custom implants are made with lattice structures that reduce material use and help with healing. In consumer products, designers are using material-efficient designs in furniture, electronics, and packaging to reduce material use without affecting strength or appearance [37].

To further improve material efficiency, manufacturers need to consider the whole life cycle of a product. This includes how it's made, used, repaired, and eventually recycled. Good design can make a product easier to repair or upgrade. Using fewer material types makes it easier to recycle. Modular design and eco-design principles also help extend product life and reduce waste. These practices help make the most of available resources and support more sustainable production.

Material efficiency depends not only on using less material but also on using it in smarter ways. Additive manufacturing, along with proper design strategies, gives engineers the tools they need to produce high-quality parts with lower resource use. This supports both environmental responsibility and efficient production. The combination of eco-friendly, performance-based design and additive manufacturing creates a clear path towards better material use and sustainable manufacturing [20].

2.5 ECO-FRIENDLY FILAMENTS AND POWDERS

The selection of suitable filaments and powders plays a pivotal role in the success of additive manufacturing (AM) operations. It directly influences not only the mechanical performance and reliability of the final product but also the broader goal of operational sustainability. Material choice is more than a technical requirement; it is a strategic decision that integrates environmental responsibility and production efficiency.

Choosing the right materials for AM requires careful evaluation of various criteria, including material composition, ease of recyclability, and associated environmental impact. As additive manufacturing continues to grow across a range of industries, the demand for eco-friendly materials has been increasing rapidly. The use of recycled or bio-based filaments and powders can reduce the environmental footprint of AM processes, which is important for production with sustainability objectives [11].

Moreover, material properties, including process parameters, such as printing temperature, layer height, and speed, affect both energy usage and material waste. By optimising these parameters, industries can reduce waste and energy consumption during the production of 3D components [31].

Figure 2.5 illustrates a comparison of sustainable materials' properties. Additive manufacturing is often known for its efficiency compared to traditional methods. The AM process reduces material waste by up to 90% relative to other subtractive manufacturing processes [20]. However, it is worth noting that the process is not completely waste-free, and the loss of material occurs, particularly from the support structures and unused powder residues, which must be managed effectively [10].

AM techniques such as direct energy deposition (DED) demonstrate economic and environmental benefits by extending tooling life and reducing the time to manufacture [8]. These techniques show how important materials selection is, as it is not only high-performing but also compatible with reuse and recycling frameworks.

Nowadays, recycled materials are becoming increasingly viable and widely adopted in AM applications [20]. Those materials are polymers such as polylactic acid (PLA), acrylonitrile butadiene styrene (ABS), and polypropylene (PP), which are gaining importance due to their ability to be reprocessed with minimal performance loss [20]. However, incorporating recycled materials requires important assessments, which include contamination, degradation of polymer chains, and loss of mechanical integrity.

Innovative approaches, such as the selective synthesis of metal alloys designed for reusability, are enabling new pathways to sustainable fabrication [38]. Simultaneously, the shift towards bio-based and biodegradable materials, including cellulose composites and advanced biopolymers such as PLA, provides an eco-friendly alternative to conventional, petroleum-derived inputs [39]. These materials are derived from renewable feedstocks and exhibit promising properties, including controlled biodegradability under specific environmental conditions.

Nonetheless, their adoption must be tempered by thorough life cycle assessments (LCAs) that consider broader impacts such as land use, water demand, and greenhouse gas emissions. Additives used in base polymers should also be examined for

Material Property Comparison

Characteristic	ABS	Nylon 12	PLA	rPETG	Bio-Resin
Tensile Strength (MPa)	40	48	65	50	45
Young's Modulus (GPa)	2.3	1.8	3.5	2.1	2.0
Impact Strength (J/m)	200	110	16	70	25
HDT @0.45MPa (°C)	98	140	55	75	45
Density (g/cm³)	1.04	1.01	1.24	1.23	1.12
Embodied Energy (MJ/kg)	95	150	54	35	65
CO2 Footprint (kg/kg)	3.8	6.2	2.7	1.5	3.0
Biodegradability	No	No	Yes\	Partial	Yes
Recyclability	Limited	Limited	Limited	High	Low
Toxicity (VOCs)	High	Medium	Low	Low	Very Low
Print Temp (°C)	220-250	240-260	190-220	220-240	180-200

FIGURE 2.5 A representation of materials property comparison of sustainable materials.

their potential to influence material properties, processing behaviour, and overall sustainability [40].

From a process perspective, optimisation is equally vital. The successful use of filaments and powders relies on matching process parameters to material characteristics. Characterisation techniques such as spectroscopy and chromatography are important techniques for identifying contaminants and ensuring the consistency of recycled materials. These are further supported by quality assurance practices, which include mechanical testing, in-process monitoring, and dimensional inspection.

Selecting sustainable materials for AM also requires a shift in mindset, from viewing waste as inevitable to managing it through intelligent resource strategies. Providing characteristics of recycled powders, such as melt flow index (MFI), reuse cycles, and traceability enhancements, is emerging as the best practice for material stewardship [10].

Looking ahead, the industry is trending towards higher mechanical performance, lower production costs, and increasingly tailored materials. This includes customised filaments engineered for specific applications, with environmental compatibility, processability, and functional performance. These innovations will continue to redefine material selection strategies, making sustainability a core principle of additive manufacturing.

Filaments of different recyclable materials are commercially available.

1. Polymer Filaments: Polymer filament characteristics, such as MFI, glass transition temperature, and tensile strength, play a vital role in their performance index. MFI determines how easily a material can be extruded during fused deposition modelling (FDM). The thermal stability is an important parameter for ensuring post-processing behaviour [41–43].
2. Additives and Composites: Additives such as carbon fibres, graphene, or natural fillers (e.g. wood particles) are often used to improve the functionality of the components. These additives and composites improve mechanical properties and influence thermal conductivity and visual appearance. However, additives must be carefully chosen as they can affect the recyclability and processability of the base polymer.
3. Recycled Polymer: Using recycled polymers is a sustainable approach, yet they pose challenges like chain scission, thermal degradation, and impurities. These issues lower mechanical performance. Chain extenders during reprocessing can partially restore the molecular integrity of recycled polymers.

Importantly, filtration and purification processes also make material suitable for use in AM [10, 44].

i. Bio-Based Polymer: Materials such as polylactic acid (PLA) are widely adopted due to their bio-based origin. However, PLA are not suitable for low-temperature applications. Polyhydroxyalkanoates (PHAs), like alternative materials, can be used as they provide better thermal resistance [45–47].
ii. Composite Filaments: When reinforcing polymers with materials like carbon fibre, glass fibre, or metal powders, significant improvements in strength, stiffness, and functional behaviour can be achieved. Aspect ratio, dispersion quality, interface adhesion between the matrix and filler, and the presence of surface-treated fibres or particles are the performance indicators of these materials [48–50].
iii. Multi-Material Printing: Multi-material printing enables the production of components with spatially varying properties, such as stiffness, colour, or electrical conductivity. This process demands material compatibility checks, particularly melting temperatures, thermal expansion, and post-processing treatments. Further, functionally graded materials and novel geometries in these materials enhance the design possibilities in AM [20, 51].
iv. Design Integration: Successfully integrating multiple materials into a single part requires adjusting the design methodology to account for thermal and mechanical compatibility. Therefore, novel designing and disassembly, including recyclability of these materials, should be considered to meet sustainability goals [16, 51].

v. Optimisation Techniques: Efficient material use is essential. Advanced methods such as Design of Experiments (DoE) help identify optimal parameter settings. Additionally, AI-driven algorithms, including genetic algorithms and particle swarm optimisation, are increasingly used to fine-tune process parameters, material compositions, and geometries to enhance both performance and sustainability [10, 11, 20].

2.5.1 Metal Powders and Additive Manufacturing

Emerging technologies, such as big data analytics, cyber-physical systems, embedded sensors, and intelligent robotics, are becoming necessary for the manufacturing environments [52–55]. Within this framework, additive manufacturing plays a pivotal role, enabling rapid and efficient modifications to digital models in response to customer demands [56]. In those fabrication techniques, fully functional components can be built in a single step, and AM reduces raw material waste while enabling complex, high-precision geometries that were previously not possible [57].

Powder Characteristics: In powder-based metal AM technologies such as selective laser melting (SLM) and electron beam melting (EBM), powder properties influence the success of the process. Important factors for developing good components are particle size distribution (PSD), morphology, and particle flowability. A narrow PSD helps improve packing density and surface smoothness, while spherical particles generally support better flow and more consistent laser interaction. Irregular particles, by contrast, can contribute to clogging and defects [16].

Alloy Selection: The alloy selection largely depends on its usability. Aluminium alloys are lightweight and a good choice for aerospace applications. Titanium alloys are suitable for medical implants. Stainless steel has good mechanical properties and excellent corrosion resistance. Nickel-based superalloys and refractory metals are preferred for harsh environments, as they are built to withstand harsh conditions without losing their structural integrity.

Recycled Metal Powders: Recycling metal powders is a circular strategy, and in powder bed fusion (PBF) and other advanced additive manufacturing processes, it has become an attractive approach. However, each reusable cycle can alter the powder's microstructure, chemical composition, and flow properties.

Therefore, techniques such as vacuum gas atomisation, plasma spheroidisation, and magnetic separation are employed to ensure the required powder quality. Further, handling powders in inert atmospheres is important to avoid oxidation and maintain consistent quality [20]. Figure 2.6 illustrates the comparison of optimisation techniques.

Process Parameter Optimisation: A successful metal AM build relies heavily on optimised process parameters. They include laser power, scan speed, layer thickness, and bed temperature. Any change in these parameters can cause defects such as porosity or warping. These defects are not good for fabricating robust parts. In-situ

Comparison of Optimisation Techniques

Characteristic	Topology Optimization	Generative Design	Lattice Structures
Material Savings	40-75%	50-85%	60-90%
Strength/Weight Gain	Up to 3x	2-5x	8-12x stiffness/kg
Key Applications	Load-bearing structures	Complex assemblies	Lightweighting Part Consolidation

FIGURE 2.6 Comparison of optimisation techniques in terms of material savings and other key applications.

monitoring techniques and feedstock control systems may help maintain consistency. Therefore, incorporating such advances may improve product quality. Further, design of experiments, genetic algorithms, particle morphology, and process parameter optimisation techniques are increasingly used to fine-tune process parameters for high-quality printed parts [11].

Role of Additives: The additive in the powder is an important strategy. The alloying elements or ceramic nanoparticles can improve the performance of the parts. For example, grain refiners enhance mechanical strength, while ceramic nanoparticles increase wear resistance. Therefore, proper selection and dispersion of additives play an important role in ensuring desired outcomes [58]. The need for uniform nanoparticle dispersion is important. Concentrated mixtures of additives help maintain consistent distribution in the powder and reduce variability in mechanical properties [16, 20].

Limitations in Laser Powder Bed Fusion: In laser powder bed fusion (LPBF), each layer of metal powder is selectively melted and solidified before the next layer is applied. This layering continues until the final 3D part is formed [59, 60–62]. Similarly, EBM builds parts using an electron beam in a vacuum environment, benefiting from high energy efficiency and rapid beam movement [63, 64]. Directed energy deposition (DED), on the other hand, feeds powder directly into the melt pool, offering benefits in terms of flexibility, material repair, and on-site manufacturing in sectors like aerospace and defence [11, 20, 65]. Although LPBF is one of the commonly used techniques in metal AM, it comes with challenges. The technique features include high energy usage, slow build speeds, and the potential for process-induced defects such as porosity, shrinkage, and residual thermal stress. These issues not only affect the quality of the fabricated parts but also impact production efficiency and cost-effectiveness of the parts [16, 66]. Therefore, selecting sustainable materials is a cornerstone in additive manufacturing for operational sustainability. The steps include increasing the usability of recycled or recyclable metals and evaluating the environmental impact of materials using LCA [2, 20, 67].

Additive Control and Environmental Impact: It is important to control the additives in the powder and filaments to ensure operational and environmental sustainability. Importantly, their use should be carefully considered. Additives like plasticisers (for flexibility), UV stabilisers (for durability), and flow enhancers can significantly impact performance. Further, there is a need for understanding the environmental impact. Further, bio-based or biodegradable additives offer more sustainable alternatives [58, 68].

Characterisation Techniques: Advanced techniques are essential for ensuring powder and filament quality. Differential scanning calorimetry (DSC) and Thermogravimetric analysis (TGA) are used for thermal analysis of filaments. Powder characteristics are important and can be characterised using standard techniques such as laser diffraction (for PSD), scanning electron microscopy (SEM), and gas pycnometer (for density). Flowability is often assessed using the Hall flowmeter test [17].

Automation and Digitisation: Importantly, operational sustainability in AM is also achieved through automation and digitisation. The automation can be a part of smart AM systems. It can optimise part orientation, digital slicing, and support generation to minimise the indirect energy use and material waste. These automation tools can contribute to enhancing the efficiency of AM processes and reducing the environmental burden associated with them [16, 17] (Tables 2.1 and 2.2).

TABLE 2.1
Evaluation of process compatibility and environmental impact.

Material	AM Process	Process Compatibility	Environmental Impact
PLA	Fused deposition modelling (FDM)	Low printing temperature, minimal warping	Biodegradable under specific conditions, lower carbon footprint
Recycled PETG	FDM	Good strength, chemical resistance	Reduces plastic waste, recyclable
Biodegradable Resins	Stereolithography and digital light processing (SLA/DLP)	High resolution, can be brittle	Potential for biodegradability, some resins may be toxic
Metal Powders	Laser powder bed fusion, electron beam melting, and direct energy deposition (LPBF/EBM/DED)	High strength, durable	Recycling reduces the need for virgin materials and high energy consumption during production

TABLE 2.2
Comparison of common versus eco-friendly options.

Feature	Common Materials	Eco-Friendly Materials
Raw Materials	Petroleum-based, virgin materials	Renewable, recycled, or biodegradable
Energy Consumption	High	Lower
Waste Generation	High	Reduced
End-of-Life	Non-biodegradable, landfill	Biodegradable, recyclable, or compostable
Applications	Wide range	More limited, depending on material properties

2.6 CONCLUSION

In conclusion, material selection is a foundational aspect of sustainable product development in additive manufacturing. By addressing critical questions about sourcing, end-of-life impact, and environmental cost, and by recognising the global drivers of change, industries can adopt more responsible practices. Life cycle assessment and thoughtful design are key tools that support this transition and promote sustainability at every stage of the manufacturing process.

REFERENCES

1. G. Prasad, H. Arunav, S. S. Dwight, M. Ghosh, A. Jayadev, & D. I. Nair, Advancing sustainable practices in additive manufacturing: A comprehensive review on material waste recyclability. *Sustainability*, 16 (2024) 10246. https://doi.org/10.3390/su162310246.
2. F. Fruggiero, A. Lambiase, R. Bonito, & M. Fera, The load of sustainability for additive manufacturing processes. *Procedia Manufacturing*, 41 (2019) 375. https://doi.org/10.1016/j.promfg.2019.09.022.
3. A. Jahan, & K. L. Edwards, VIKOR method for material selection problems with interval numbers and target-based criteria. *Materials & Design (1980–2015)*, 47 (2013) 759. https://doi.org/10.1016/j.matdes.2012.12.072.
4. R. Morrow, M. Ribul, H. Eastmond, A. Lanot, & S. Baurley, Bio-producing bacterial cellulose filaments through co-designing with biological characteristics. *Materials*, 16 (2023) 4893. https://doi.org/10.3390/ma16144893.
5. K. Kellens, R. Mertens, D. Paraskevas, W. Dewulf, & J. R. Duflou, Environmental impact of additive manufacturing processes: Does AM contribute to a more sustainable way of part manufacturing? *Procedia CIRP*, 61 (2017) 582. https://doi.org/10.1016/j.procir.2016.11.153.
6. P. Saxena, P. Stavropoulos, J. D. Kechagias, & K. Salonitis, Sustainability assessment for manufacturing operations. *Energies*, 13 (2020) 2730. https://doi.org/10.3390/en13112730.
7. S. Li, (2024). Sustainable material selection in architectural design. In Y. Chen et al. (Eds.), *Proceedings of the 3rd International Conference on Culture, Design and Social*

Development (CDSD 2023) (Advances in Social Science, Education and Humanities Research, Vol. 834). Atlantis Press. https://doi.org/10.2991/978-2-38476-222-4_2.

8. A. Santanaa, P. Afonsoa, A. Zaninb, & R. Wernkeb, Costing models for capacity optimization in Industry 4.0: Trade-off between used capacity and operational efficiency. *Procedia Manufacturing*, 13 (2017) 1183–1190.
9. I. Ribeiro, F. Matos, C. Jacinto, H. Salman, G. Cardeal, H. Carvalho, R. Godina, & P. Peças, Article framework for life cycle sustainability assessment of additive manufacturing. *Sustainability*, 12(3) (2020) 929.
10. T. Peng, K. Kellens, R. Tang, C. Chen, & G. Chen, Sustainability of additive manufacturing: An overview on its energy demand and environmental impact. *Additive Manufacturing*, 21 (2018) 694. https://doi.org/10.1016/j.addma.2018.04.022.
11. M. Gopal, H. G. Lemu, & E. M. Gutema, Sustainable additive manufacturing and environmental implications: Literature review. *Sustainability*, 15 (2022) 504. https://doi.org/10.3390/su15010504.
12. T. M. Tavares, M. G. Filho, G. M. D. Ganga, & M. H. B. M. Callefi, The relationship between additive manufacturing and circular economy: A sistematic review. *Independent Journal of Management & Production*, 11 (2020) 1648. https://doi.org/10.14807/ijmp.v11i5.1290.
13. M. Sauerwein, E. L. Doubrovski, R. Balkenende, & C. Bakker, Exploring the potential of additive manufacturing for product design in a circular economy. *Journal of Cleaner Production*, 226 (2019) 1138. https://doi.org/10.1016/j.jclepro.2019.04.108.
14. R. Cioffi, M. Travaglioni, G. Piscitelli, A. Petrillo, & A. Parmentola, Smart manufacturing systems and applied industrial technologies for a sustainable industry: A systematic literature review. *Applied Sciences*, 10 (2020) 2897. https://doi.org/10.3390/app10082897.
15. F. L. Garcia, V. A. da S. Moris, A. O. Nunes, & D. A. L. Silva, Environmental performance of additive manufacturing process – an overview. *Rapid Prototyping Journal,* 24 (2018) 1166. https://doi.org/10.1108/rpj-05-2017-0108.
16. P. Nyamekye, R. Lakshmanan, V. Tepponen, & S. Westman, Sustainability aspects of additive manufacturing: Leveraging resource efficiency via product design optimization and laser powder bed fusion. *Heliyon*, 10(1) (2023) e23152. https://doi.org/10.1016/j.heliyon.2023.e23152.
17. C. G. Mançanares, E. Zancul, & P. A. C. Miguel, Sustainable manufacturing strategies: A literature review on additive manufacturing approach. *Product Management & Development*, 13 (2015) 47. https://doi.org/10.4322/pmd.2015.001.
18. Y. Tang, K. Mak, & Y. F. Zhao, A framework to reduce product environmental impact through design optimization for additive manufacturing. *Journal of Cleaner Production*, 137 (2016) 1560. https://doi.org/10.1016/j.jclepro.2016.06.037.
19. G. Taddese, S. Durieux, & E. Duc, Sustainability performance indicators for additive manufacturing: A literature review based on product life cycle studies. *The International Journal of Advanced Manufacturing Technology*, 107 (2020) 3109. https://doi.org/10.1007/s00170-020-05249-2.
20. H. Hegab, N. Khanna, N. Monib, & A. Salem, Design for sustainable additive manufacturing: A review. *Sustainable Materials and Technologies*, 35 (2023) e00576. https://doi.org/10.1016/j.susmat.2023.e00576.
21. R. Godina, I. Ribeiro, F. Matos, B. T. Ferreira, H. Carvalho, & P. Peças, Article impact assessment of additive manufacturing on sustainable business models in industry 4.0 context. *Sustainability*, 12(17) (2020) 7066.
22. J.-J. Kim & B. Rigdon, (1998). *Qualities, use, and examples of sustainable building materials. Sustainable Architecture Module* (National Pollution Prevention Center for

Higher Education, College of Architecture and Urban Planning, The University of Michigan, Ann Arbor, MI, USA) pp. 1–44.

23. H. A. Colorado, E. I. G. Velásquez, & S. N. Monteiro, Sustainability of additive manufacturing: The circular economy of materials and environmental perspectives. *Journal of Materials Research and Technology*, 9 (2020) 8221. https://doi.org/10.1016/j.jmrt.2020.04.062.
24. M. K. Niaki, S. A. Torabi, & F. Nonino, Why manufacturers adopt additive manufacturing technologies: The role of sustainability. *Journal of Cleaner Production*, 222 (2019) 381. https://doi.org/10.1016/j.jclepro.2019.03.019.
25. S. Choe, Y. Kim, Y. Won, & J. Myung, Bridging three gaps in biodegradable plastics: Misconceptions and truths about biodegradation. *Frontiers in Chemistry*, 9 (2021) 671750. https://doi.org/10.3389/fchem.2021.671750.
26. K. Ganesan, B. Guin, E. Wilbanks, & J. C. Sternberg, Synthesis and characterization of soy hull biochar-based flexible polyurethane foam composites. *Materials*, 18 (2025) 2006. https://doi.org/10.3390/ma18092006.
27. L. W. McKeen, *Renewable Resource and Biodegradable Polymers*. Elsevier eBooks (Elsevier BV, 2012), p. 305. https://doi.org/10.1016/b978-1-4557-2598-4.00012-5.
28. M. Tahir, & A. M. Seyam, Greening fused deposition modeling: A critical review of plant fiber-reinforced PLA-based 3D-printed biocomposites. *Fibers*, 13 (2025) 64. https://doi.org/10.3390/fib13050064.
29. J. Rydz, W. Sikorska, M. Musioł, B. Zawidlak-Węgrzyńska, & K. Duale, *Sustainable Future Alternative: (Bio)degradable Polymers for the Environment*. Elsevier eBooks (Elsevier BV, 2019), p. 274. https://doi.org/10.1016/b978-0-12-803581-8.11520-6.
30. K. Hom, Circular economy: The 4Rs reduce-reuse-recycle-recover. In *IOP Conference Series Earth and Environmental Science* (IOP Publishing, 2024), p. 12042. https://doi.org/10.1088/1755-1315/1402/1/012042IOP. Conference Series: Earth and Environmental Science, Volume 1402, International Conference on Challenges for the Next Generation Built Environment 09/05/2024–10/05/2024, Bologna, Italy.
31. S. Hauck, L. Greif, N. Benner, & J. Ovtcharova, Advancing sustainable additive manufacturing: Analyzing parameter influences and machine learning approaches for CO2 prediction. *Sustainability*, 17 (2025) 3804. https://doi.org/10.3390/su17093804.
32. M. Javaid, A. Haleem, R. P. Singh, R. Suman, & S. Rab, Role of additive manufacturing applications towards environmental sustainability. *Advanced Industrial and Engineering Polymer Research*, 4 (2021) 312. https://doi.org/10.1016/j.aiepr.2021.07.005.
33. M. Kumaran, & V. Senthilkumar, Generative design and topology optimization of analysis and repair work of industrial robot arm manufactured using additive manufacturing technology. In *IOP Conference Series Materials Science and Engineering* (IOP Publishing, 2021), p. 12036. https://doi.org/10.1088/1757-899x/1012/1/012036.
34. M. S. Martins, R. B. O. Acevedo, S. Glyniadakis, H. P. Machado, F. C. Albuquerque, K. S. Ferreira, Y. L. A. Benazio, & E. C. Medeiros, Topological optimization and generative design for additive manufacturing: Software, trends, and opportunities. *ASME Journal of Mechanical Design* 148(1) (2026) 014501. https://doi.org/10.1115/1.4068852.
35. I. Ntintakis, & G. E. Stavroulakis, Progress and recent trends in generative design. *MATEC Web of Conferences*, 318 (2020) 1006. https://doi.org/10.1051/matecconf/202031801006.
36. J. Coulthard, & C. Wang, Generative design and topology optimisation of products for additive manufacturing. In *Advances in Transdisciplinary Engineering* (IOS Press, 2022). https://doi.org/10.3233/atde220570.

37. A. Aabid, & M. Baig, Sustainable materials for engineering applications. *Materials*, 16 (2023) 6085. https://doi.org/10.3390/ma16186085.
38. V. Sahajwalla, & R. Hossain, Rethinking circular economy for electronics, energy storage, and solar photovoltaics with long product life cycles. *MRS Bulletin*, 48 (2023) 375. https://doi.org/10.1557/s43577-023-00519-2.
39. A. D. Macheca, B. K. Mutuma, J. L. Adalima, E. Midheme, L. H. M. Lucas, V. K. Ochanda, & S. D. Mhlanga, Perspectives on plastic waste management: Challenges and possible solutions to ensure its sustainable use. *Recycling*, 9 (2024) 77. https://doi.org/10.3390/recycling9050077.
40. I.-C. Radu, A.-M. Vadureanu, D. Cozorici, E. Blanzeanu, & C. Zaharia, Advancing sustainability in modern polymer processing: Strategies for waste resource recovery and circular economy integration. *Polymers*, 17 (2025) 522. https://doi.org/10.3390/polym17040522.
41. V. N. Andronov, L. Beránek, V. Krůta, L. Hlavůňková, & Z. Jeníková, Overview and comparison of PLA filaments commercially available in Europe for FFF technology. *Polymers*, 15 (2023) 3065. https://doi.org/10.3390/polym15143065.
42. W. Wu, W. Ye, P. Geng, Y. Wang, G. Li, X. Hu, & J. Zhao, 3D printing of thermoplastic PI and interlayer bonding evaluation. *Materials Letters*, 229 (2018) 206. https://doi.org/10.1016/j.matlet.2018.07.020.
43. T. Ma, Y. Zhang, K. Ruan, H. Guo, M. He, X. Shi, Y. Guo, J. Kong, & J. Gu, Advances in 3D printing for polymer composites: A review. *InfoMat*, 6 (2024) e12568. https://doi.org/10.1002/inf2.12568.
44. F. Knappich, M. Klotz, M. Schlummer, J. Wölling, & A. Mäurer, Recycling process for carbon fiber reinforced plastics with polyamide 6, polyurethane and epoxy matrix by gentle solvent treatment. *Waste Management*, 85 (2018) 73. https://doi.org/10.1016/j.wasman.2018.12.016.
45. N. Peelman, P. Ragaert, K. Ragaert, B. D. Meulenaer, F. Devlieghere, & L. Cardon, Heat resistance of new biobased polymeric materials, focusing on starch, cellulose, PLA, and PHA. *Journal of Applied Polymer Science*, 132 (2015) 42305 (1–15). https://doi.org/10.1002/app.42305.
46. R. Balart, N. M. Muñoz, F. Dominici, T. Boronat, & S. Torres-Giner, Environmentally friendly polymers and polymer composites. *Materials*, 13 (2020) 4892. https://doi.org/10.3390/ma13214892.
47. M. B. Подзорова, Yu. V. Tertyshnaya, С. Г. Карпова, & А. А. Попов, Impact of UV treatment on polylactide–polyethylene film properties. *IOP Conference Series Materials Science and Engineering*, 525 (2019) 12043. https://doi.org/10.1088/1757-899x/525/1/012043.
48. P. K. Vallittu, High-aspect ratio fillers: Fiber-reinforced composites and their anisotropic properties. *Dental Materials*, 31 (2014) 1. https://doi.org/10.1016/j.dental.2014.07.009.
49. M. N. Norizan, F. A. Sabaruddin, M. M. Harussani, S. H. Kamarudin, M. Rayung, M. R. M. Asyraf, H. A. Aisyah, M. N. F. Norrrahim, R. A. Ilyas, N. Abdullah, E. S. Zainudin, S. M. Sapuan, & K. Abdan, Mechanical performance and applications of CNTs reinforced polymer composites – A review. *Nanomaterials,* 11 (2021) 2186. https://doi.org/10.3390/nano11092186.
50. C. Harito, D. V. Bavykin, B. Yuliarto, H. K. Dipojono, & F. C. Walsh, Polymer nanocomposites having a high filler content: Synthesis, structures, properties, and applications. *Nanoscale*, 11 (2019) 4653. https://doi.org/10.1039/c9nr00117d.
51. H. Salem, H. Abouchadi, & K. El Bikri, Design for additive manufacturing. *Journal of Theoretical and Applied Information Technology*, 98(19) (2020) 3042–3054.

52. C. Panwisawas, Y. T. Tang, & R. C. Reed, Metal 3D printing as a disruptive technology for superalloys. *Nature Communications*, 11 (2020) 2327. https://doi.org/10.1038/s41467-020-16188-7.
53. J. P. Manaia, F. Cerejo, & J. Duarte, Revolutionising textile manufacturing: A comprehensive review on 3D and 4D printing technologies. *Fashion and Textiles*, 10 (2023) 20. https://doi.org/10.1186/s40691-023-00339-7.
54. D. Mladenovska, A. M. Lazarevska, & M. Krstanoski, Attributes relevant for sustainable additive manufacturing – material driven approach. In *IOP Conference Series Materials Science and Engineering* (IOP Publishing, 2021), p. 12001. https://doi.org/10.1088/1757-899x/1190/1/012001.
55. G. M. Cappucci, M. Pini, P. Neri, M. Marassi, E. Bassoli, & A. M. Ferrari, Environmental sustainability of orthopedic devices produced with powder bed fusion. *Journal of Industrial Ecology*, 24 (2019) 681. https://doi.org/10.1111/jiec.12968.
56. M. Vorkapić, M. A. Hasan, D. Tanović, M. Baltić, & B. Tomić, Implementation of 6R strategy in FDM printing process: Case: Small electronic enclosure box. *Journal of Engineering Management and Competitiveness*, 10 (2020) 141. https://doi.org/10.5937/jemc2002141v.
57. B. Vayre, F. Vignat, & F. Villeneuve, Metallic additive manufacturing: State-of-the-art review and prospects. *Mechanics & Industry*, 13 (2012) 89. https://doi.org/10.1051/meca/2012003.
58. R. Höfer, & K. Hinrichs, Additives for the manufacture and processing of polymers. In P. Eyerer, M. Weller, & C. Hübner (Eds.), *Polymers – Opportunities and Risks II. The Handbook of Environmental Chemistry* (Springer Nature, 2009) Vol 12, p. 97. https://doi.org/10.1007/698_2009_12.
59. V. Thampy, A. Y. Fong, N. P. Calta, J. Wang, A. A. Martin, P. J. Depond, A. M. Kiss, G. Guss, Q. Xing, R. Ott, T. van Buuren, M. F. Toney, J. N. Weker, M. J. Kramer, M. J. Matthews, C. J. Tassone, & K. H. Stone, Subsurface cooling rates and microstructural response during laser based metal additive manufacturing. *Scientific Reports*, 10 (2020) 1981. https://doi.org/10.1038/s41598-020-58598-z.
60. G. Boussinot, M. Apel, J. Zielinski, U. Hecht, & J. H. Schleifenbaum, Strongly out-of-equilibrium columnar solidification during laser powder-bed fusion in additive manufacturing. *Physical Review Applied*, 11 (2019) 014025. https://doi.org/10.1103/physrevapplied.11.014025.
61. A. Peles, V. Paquit, & R. Dehoff, Deep-learning quantitative structural characterization in additive manufacturing. arXiv (Cornell University) (2023). https://doi.org/10.48550/arxiv.2302.06389.
62. K. Zhang, X. M. Zhang, & X. F. Shang, Research on cladding process of metal powder during laser additive manufacturing. *Applied Mechanics and Materials*, 380 (2013) 4311. https://doi.org/10.4028/www.scientific.net/amm.380-384.4311.
63. M. F. Zaeh, S. Lutzmann, G. Branner, & G. F. Strasser, Solutions for modelling the energy input in electron beam material processing. *Engineering Systems Design and Analysis*, 48357 (2008) 459–467. https://doi.org/10.1115/esda2008-59539.
64. V. Perumal, A. R. Najafi, & A. Kontsos, A novel digital design approach for metal additive manufacturing to address local thermal effects. *Designs*, 4 (2020) 41. https://doi.org/10.3390/designs4040041.
65. L. Scime, & J. Beuth, Anomaly detection and classification in a laser powder bed additive manufacturing process using a trained computer vision algorithm. *Additive Manufacturing*, 19 (2017) 114. https://doi.org/10.1016/j.addma.2017.11.009.
66. W. Abd-Elaziem, S. Elkatatny, T. A. Sebaey, M. A. Darwish, M. A. A. El-baky, & A. Hamada, Machine learning for advancing laser powder bed fusion of stainless

steel. *Journal of Materials Research and Technology*, 30 (2024) 4986. https://doi.org/10.1016/j.jmrt.2024.04.130.

67. R. Rahmani, B. Bashiri, S. I. Lopes, A. Hussain, H. S. Maurya, & R. Vilu, Sustainable additive manufacturing: An overview on life cycle impacts and cost efficiency of laser powder bed fusion. *Journal of Manufacturing and Materials Processing*, 9 (2025) 18. https://doi.org/10.3390/jmmp9010018.
68. R. R. A. Silva, C. S. Marques, T. R. Arruda, S. C. Teixeira, & T. V. de Oliveira, Biodegradation of polymers: Stages, measurement, standards and prospects. *Macromol – A Journal of Macromolecular Research*, 3 (2023) 371. https://doi.org/10.3390/macromol3020023.

3 Product Design for Operational Sustainability

3.1 INTRODUCTION

Operational sustainability is not attained solely through material selections or manufacturing efficiency. It fundamentally begins at the design stage. Product design profoundly influences sustainability by determining a product's entire life cycle. This includes material selection, manufacturing processes, distribution, usage, and end-of-life management [1]. Considering that a product's development phase is responsible for a significant portion of its environmental impact, as much as 80%, corporations are adopting methods like design for the environment, recycling, reuse, and remanufacturing [2]. Sustainable product design integrates environmental considerations alongside traditional factors like performance, aesthetics, durability, and cost [3].

The integration of sustainability principles into product design necessitates a holistic approach, which encompasses the entire product lifecycle [4]. It involves making informed decisions about materials, manufacturing processes, energy consumption during use, and end-of-life management. Further, it supports reducing environmental impact and increasing product value [5]. The right design can extend a product's life cycle. The reuse, refurbishment, or remanufacturing of products can reduce manufacturing-related environmental impacts and the need for new materials when products are designed properly [4]. Usually, the design process includes disassembly and material recovery to encourage recycling and waste reduction [6].

Achieving operational sustainability requires that sustainability principles be incorporated into product design. Therefore, a comprehensive strategy that takes into account every stage of the product lifecycle, from the extraction of raw materials to the management of end-of-life, is required [7]. Planning for end-of-life recovery, reducing resource consumption, and increasing product lifespan all start with product design. This chapter explores the role that design plays in supporting sustainable manufacturing, with a focus on digital design tools, additive manufacturing (AM), and circular economy strategies. Frequent replacement of the components in service can be reduced by durable and robust product designs. This step can also reduce resource consumption and environmental effects overall.

This chapter defines and critically explores the role that product design can play in supporting sustainable manufacturing. It provides particular attention to digital

DOI: 10.1201/9781003633754-3

product design tools, additive manufacturing processes, and circular economy strategies. Further, it looks at how design can improve product longevity, cutting down on resource usage and preparing for end-of-life recovery. Therefore, by focusing on design, a solid foundation for operational sustainability across the entire product lifecycle can be created.

3.2 DESIGN FOR ADDITIVE MANUFACTURING

The approach for product design is offered by additive manufacturing. Further, the outcome of this approach is operational sustainability that distinguishes it from the subtractive manufacturing techniques [8–9].

Additive manufacturing builds parts layer by layer. It reduces material waste. It also makes complex shapes possible. These shapes make better use of material. This is different from subtractive methods, which cut away material.

Design for Additive Manufacturing means designing products to get the most from 3D printing. This lets us create parts that are lighter, stronger, and more efficient. Additive manufacturing can create complex shapes and use different materials. It has useful features. It emphasises what works best, not merely what can be made [10].

Since it builds parts layer by layer, it helps us use less material, making the process more efficient and environmentally friendly. There is less waste. Designs can be changed to use only what is needed. This saves resources. It also allows the creation of more complex shapes. These shapes are hard or impossible to make with traditional methods [11]. It supports on-demand manufacturing. This reduces the need for bulk storage and cuts waste from unsold or outdated stock.

It reduces waste and makes processes cleaner. It also promotes the use of new, innovative materials [12]. Products can be customised for specific needs. This boosts user satisfaction and helps prevent waste from overproduction or products becoming outdated. Specific needs, which not only improve user satisfaction but also help prevent waste from overproduction or products becoming obsolete.

If just one part of a product breaks, it can be reprinted instead of replacing the entire item [13]. On-demand printing also reduces the need for large inventories, resulting in lower storage and waste [14].

The aerospace industry is particularly benefiting from this technology. It can produce strong, lightweight parts directly from digital designs. This enables engineers to use advanced tools such as topology optimisation to reduce weight and enhance fuel efficiency [15].

Complex features such as internal cooling channels can be integrated directly into parts, improving performance [16]. Additive manufacturing simplifies the design of multi-material, multifunctional, and highly customised components, helping us use resources more efficiently and protect our environment [17]. Topology optimisation and generative design are innovative tools used to minimise the amount of material in a product [12].

These approaches result in designs that are both structurally efficient and lightweight. By integrating additive manufacturing technologies at the enterprise level, distance and transportation needs are expected to decrease, as manufacturing can be closer to consumption, thereby reducing negative environmental impacts [2]. Unlike conventional manufacturing, which often restricts design due to subtractive limitations, additive manufacturing allows far greater flexibility. Engineers can design complex geometries, internal channels, and customised structures that are difficult with traditional methods.

Several key design strategies help maximise the benefits of AM. Reducing support structures cuts material waste and shortens post-processing time. Embedding functional features, like hinges, clips, or fasteners, directly into the printed part can eliminate the need for secondary components. Part consolidation reduces assembly steps and fastener use, improving reliability and lowering material consumption. Lightweighting techniques, such as lattice infill and topology optimisation, further reduce weight while maintaining strength. By designing specifically for AM, engineers can achieve higher performance, lower material use, and more sustainable outcomes.

3.3 GENERATIVE DESIGN AND TOPOLOGICAL OPTIMISATION

Generative design and topology optimisation are powerful digital techniques that enable engineers to reconsider how products are shaped. Both depend on computational algorithms, but their goals are clear: to produce lighter, stronger, and more efficient components.

Topology optimisation focuses on removing material from regions of a component that experience little or no stress. The result is a lighter structure that still meets performance demands. Such designs often resemble natural, organic forms rather than traditional machine parts, as highlighted by studies on lightweight structural design [14].

Generative design builds on this principle by generating multiple design options at once, each optimised for goals such as reduced weight, increased stiffness, or lower cost. These designs are not limited by traditional manufacturing rules, and many of them can only be produced using additive manufacturing. Research shows that this approach results in components that are significantly lighter and stronger while using far less material [18–20].

The benefits extend beyond the component itself. Techniques like generative design and topology optimisation have demonstrated their ability to decrease material consumption, expedite manufacturing processes, and minimise energy use, all without compromising performance or safety. Additionally, they promote sustainability goals; digital optimisation helps with waste and raw material needs, aligning economic principles.

This is evident in studies showing material and mass savings in both plastic and metal components [11, 21].

Applying topology optimisation in practice requires a clear problem definition. Engineers must specify the loads, boundary conditions, geometric limits, and material properties. Once these constraints are defined, computational methods generate optimised designs that can be validated using finite element analysis. The resulting solutions, often unconventional in appearance, are structurally efficient and reliable

[11, 22]. Generative design, in particular, has proven valuable for lightweighting, with studies reporting weight reductions of up to 50% in aerospace components such as antenna support arms, without compromising in performance [11, 23].

These methods also improve efficiency across the entire life cycle of a product. Reducing part weight through simulation-driven optimisation decreases the amount of raw material needed, shortens manufacturing time, and lowers energy use during production. Recent methods integrate computational tools with material-aware models and stochastic processes, allowing for the simultaneous optimisation of both the topology and the material structure. The use of generative design and topology optimisation introduces a new approach for engineers. When combined with 3D printing, these technologies enable the creation of complex, customised shapes that improve performance, minimise material waste, and promote more sustainable manufacturing processes [24, 25].

3.4 DESIGN AND STRATEGIES FOR END-OF-LIFE RECYCLING PROCESSES

End-of-life recycling plays a crucial role in sustainability, helping us protect our environment for future generations. It lowers environmental impact and supports the circular economy. Unlike the linear 'make–use–dispose' model, the circular approach keeps materials and products in use for longer [26–28].

Additive manufacturing can only be truly sustainable if end-of-life is part of product design. Products must be easy to take apart so that parts can be reused or recycled [17]. Material choice is important. Recyclable and non-hazardous materials work best. Modular designs are beneficial because individual parts can be replaced without discarding the entire product [29].

These strategies extend product life and reduce waste. The 6R framework – reduce, reuse, recycle, recover, redesign, and remanufacture – supports closed-loop systems [17]. Designing for repair and recovery mitigates environmental impact throughout the entire product lifecycle [26]. Applying eco-design principles early is economical and guarantees sustainability from the outset. Raw material sourcing must also be considered, as it affects repair, maintenance, and recycling [14].

Designing for end-of-life offers clear advantages. It enhances sustainability, facilitates disassembly and recycling [30, 31], prolongs product lifespan, and reduces the overall footprint. It also ensures resources are utilised more efficiently, supporting the 'triple bottom line' of environment, society, and economy [4, 17]. This requires moving from linear to circular systems. It reduces planned obsolescence and the culture of throwaway products [4, 32, 33].

Key strategies include simple disassembly, recyclable material choices, and mono-material structures. Modular parts allow easy replacement. Products should be designed to withstand use and allow reprocessing when required. Triggered obsolescence, in which materials degrade as intended, is another possibility [34]. Durability and user attachment also extend lifetimes [35, 36].

Digital manufacturing supports these ideas. Additive manufacturing allows on-demand production and reduces waste [14]. However, challenges remain. Intellectual property concerns, emissions, and full life cycle assessments continue to be issues [14, 37]. Supply chains must be responsible, ensuring recycled materials are ethically sourced [14]. Without this, sustainability cannot be achieved [26, 38].

The benefits are substantial. Products last longer through repair and reuse [14, 35, 39]. Waste and landfill loads are reduced, while recycling creates new jobs [11, 34]. Material reuse reduces emissions and lowers dependence on feedstock resources [14, 40, 41]. Companies also save costs on materials and waste management. Remanufacturing can even turn disposal costs into profit [42–45]. New recycling methods can provide competitive advantages [46]. Implementing sustainable design enhances reputation and draws in eco-conscious consumers [47–52].

There are still some challenges to overcome. Electronics and Information and Communication Technology (ICT) devices typically contain a variety of materials and use intricate fastening techniques, which can make recycling a bit tricky [14, 26, 53]. Low-durability products are even more difficult to recover and often end up discarded [54].

By emphasising smarter design, supportive policies, and innovative technologies, we can collaboratively overcome these challenges. Adopting a lifecycle perspective – covering everything from sourcing materials to recycling – enables us to make more sustainable decisions [55]. Creating products that are durable, repairable, and recyclable extends their usefulness. Additionally, using transparent materials, simplifying disassembly, and designing modular products help conserve resources [17]. This approach not only reduces waste but also promotes sustainability beyond the active use phase of the product [11].

3.5 DESIGN FOR LIGHTWEIGHT STRUCTURES

Designing for lightweight structures is a crucial part of sustainable product design. It boosts efficiency and minimises environmental impact throughout the product's lifecycle. Decreasing weight reduces energy consumption during manufacturing, transportation, and operation. This is especially important in the aerospace and automotive sectors, where lighter parts improve better fuel efficiency and reduce greenhouse gas emissions [26].

Lightweighting uses advanced materials and smart structural designs. These methods achieve high performance while reducing the impact of material extraction and processing [4, 26]. The method also aligns with circular economy principles by promoting durability and reuse [56]. The benefits of weight reduction are clear. Using less material during production saves energy. Lighter products decrease transport fuel consumption. In dynamic industries such as aerospace and automotive, lightweight components enhance efficiency during use.

However, the high energy cost of producing some advanced lightweight materials must also be considered to ensure a net environmental gain [4]. Additive manufacturing allows for complex shapes and internal geometries that reduce weight without compromising strength. Lattice structures, thin-walled parts, and hollow sections save material. Biomimetic structures inspired by nature provide excellent strength-to-weight ratios. Computational techniques like topology optimisation remove unnecessary material from low-stress areas. Generative design creates multiple options within set limits, leading to lighter, optimised forms. Together, these tools allow engineers to design for AM and achieve high performance with reduced weight.

The advantages are extensive. Material demand decreases, reducing extraction impacts. Energy consumption drops during manufacturing, transport, and use. Lightweight components enhance speed, handling, and durability, resulting in a more responsive and robust experience. They reduce costs by minimising material and energy consumption, which benefits both your budget and the environment. Furthermore, a lightweight design encourages innovation by fostering new ideas and creative solutions in design and manufacturing. Additive manufacturing is essential here, allowing the creation of complex lattice and biomimetic structures while improving material efficiency for optimal results.

Computational tools such as topology optimisation ensure materials are used only where needed [57, 58]. Combining these tools with AM produces optimised structures that improve power density, carrying capacity, and overall system efficiency [59]. The 'complexity-for-free' nature of AM also allows part consolidation and geometry optimisation, leading to significant lifecycle energy savings [60].

3.6 DESIGN FOR CIRCULAR ECONOMY

The circular economy seeks to keep materials and products in use for as long as possible. Design has a central role in achieving this. It enables repair, refurbishment, and recycling by using modular systems and recyclable materials. This extends life cycles and reduces waste [61].

Good design facilitates easy repair and upgrading. Modular components can be swapped out instead of disposing of entire products. Recycling is made easier when products use fewer mixed or composite materials. Product–service models, such as leasing or remanufacturing, also extend product life. These strategies must be built into the design stage to achieve sustainability [62].

A lifecycle approach considers extraction, manufacturing, use, and disposal. It supports strategies such as durability, repairability, and recyclability to maximise value [63]. Robust products tend to last longer and need less maintenance. Durable components can be reused in new products after their lifecycle ends, fostering sustainability and encouraging innovation. Designers can consider remanufacturing, a process that restores products to nearly new condition, which supports sustainability and prolongs product lifespan. This approach often provides greater environmental benefits than simply recycling, making it an excellent way for us to help protect our planet. This method often offers greater environmental benefits than just recycling, making it an excellent way for us to help protect our planet [64, 65].

3.7 AI AND MACHINE LEARNING IN OPERATIONAL SUSTAINABILITY FOR DESIGN

Artificial intelligence and machine learning are significantly changing the approach to sustainable design, opening up new possibilities for a more environmentally conscious future. They provide tools to predict performance, recommend environmentally friendly materials, and optimise processes, which help reduce waste and energy use while improving efficiency [66].

Designers use digital simulations to evaluate product performance. This reduces the need for physical prototypes. Recommendation systems suggest greener materials and design solutions. Real-time monitoring allows adjustments during production, reducing errors and improving resource use. AI also supports optimisation of complex processes, cutting energy and material waste [66].

In materials science, AI accelerates the discovery of new materials with enhanced sustainability [67]. Integrating AI into circular design helps industries reduce costs and improve resource efficiency [68]. AI tools also enhance geometry optimisation, improving function and efficiency [26]. Their autonomous nature reduces human error and improves reliability. They also process large datasets to identify better design and recycling strategies [69]. AI-driven analysis reduces the time and energy required for data processing and enables rapid prototyping, further reducing waste [69].

3.8 COST CALCULATION TO COMMERCIALISATION

Sustainable product design must also be economically viable. Lifecycle cost analysis (LCCA) is essential. It assesses costs from design through to disposal or recycling. This ensures decisions are made with full financial awareness. Cost-benefit models compare environmental savings with investment needs, showing the overall financial impact [11].

Scalability is still an important consideration. It's essential that sustainable designs can be produced in large volumes without high costs or increased complexity [70, 71]. Additionally, it's worth taking a closer look at the costs associated with tooling and post-processing, especially in additive manufacturing [14, 37, 72, 73]. These assessments, integrated into the design process, ensure that products are both sustainable and commercially viable. A clear financial strategy boosts the competitiveness and attractiveness of eco-friendly innovations, turning environmentally friendly ideas into successful market offerings. This approach promotes growth and economic viability without repetition.

3.9 CONCLUSION

The basis of operational sustainability relies on how a product is designed. Every decision taken during the design phase influences the choice of materials, manufacturing methods, energy consumption, and the ease of repair or recycling once the product reaches end-of-life.

A well-organised design can therefore cut waste, prolong product lifespan, and lessen environmental impact without sacrificing performance or function.

Additive manufacturing offers new opportunities to design components that are lighter, stronger, and more efficient. It permits the use of only the necessary amount of material and enables the creation of complex shapes that cannot be produced by traditional methods. Generative design and topology optimisation further enhance this process by determining the most effective use of material, resulting in weight reduction and improved performance.

Designing with recycling and reuse in mind ensures that materials can re-enter production cycles instead of becoming waste. Lightweight structures and modular designs facilitate easier repair and maintenance, supporting industries in moving towards a circular model of production and consumption.

Digital tools, such as data-driven modelling and automated design systems, have made it possible to assess cost, energy, and resource use early in the product development process. When such assessments are combined with life cycle cost analysis, designers can balance environmental responsibility with economic practicality.

In summary, sustainable product design is both a technical and ethical obligation. It skilfully melds creativity with environmental stewardship and a focus on long-term efficiency. When industries integrate sustainable practices from the outset of design, they can achieve enduring productivity, reduce dependency on resources, and make a significant contribution towards reaching operational sustainability objectives.

REFERENCES

1. K. Ramani, D. Ramanujan, W. Z. Bernstein, F. Zhao, J. W. Sutherland, C. A. Handwerker, J. Choi, H. Kim, & D. Thurston, Integrated sustainable life cycle design: A review. *Journal of Mechanical Design*, 132 (2010) 091004 (1–15). https://doi.org/10.1115/1.4002308.
2. C. G. Mançanares, E. Zancul, & P. A. C. Miguel, Sustainable manufacturing strategies: A literature review on additive manufacturing approach. *Product Management & Development*, 13 (2015) 47. https://doi.org/10.4322/pmd.2015.001.
3. J. Han, P. Jiang, & P. Childs, Metrics for measuring sustainable product design concepts. *Energies*, 14 (2021) 3469. https://doi.org/10.3390/en14123469.
4. P. Saxena, P. Stavropoulos, J. D. Kechagias, & K. Salonitis, Sustainability assessment for manufacturing operations. *Energies*, 13 (2020) 2730. https://doi.org/10.3390/en13112730.
5. B. M. Hapuwatte, & I. S. Jawahir, Closed-loop sustainable product design for circular economy. *Journal of Industrial Ecology*, 25 (2021) 1430. https://doi.org/10.1111/jiec.13154.
6. F. H. Ho, A.-R. S. Hanim, R. G. R. Ariffin, M. N. A. Rahman, M. F. Hassan, A. M. T. Arifin, R. H. A. Haq, S. Ahmad, & H. Abdullah, Analytical analysis for material efficiency practice: An exploratory study of Malaysia manufacturing cases. *Journal of Physics Conference Series*, 1150 (2019) 12006. https://doi.org/10.1088/1742-6596/1150/1/012006.
7. P. Upadhyay, & A. Kumar, A house of sustainability-based approach for green product design. *Management of Environmental Quality An International Journal*, 31 (2019) 819. https://doi.org/10.1108/meq-03-2019-0057.
8. O. Diegel, S. Singamneni, S. Reay, & A. Withell, Tools for sustainable product design: Additive manufacturing. *Journal of Sustainable Development*, 3 (2010) 68. https://doi.org/10.5539/jsd.v3n3p68.
9. M. N. Nadagouda, M. Ginn, & V. Rastogi, A review of 3D printing techniques for environmental applications. *Current Opinion in Chemical Engineering*, 28 (2020) 173. https://doi.org/10.1016/j.coche.2020.08.002.

10. F. Giudice, R. Barbagallo, & G. Fargione, A design for additive manufacturing approach based on process energy efficiency: Electron beam melted components. *Journal of Cleaner Production*, 290 (2020) 125185. https://doi.org/10.1016/j.jclepro.2020.125185.
11. A. Santanaa, P. Afonsoa, A. Zaninb, & R. Wernkeb, Costing models for capacity optimization in Industry 4.0: Trade-off between used capacity and operational efficiency. *Procedia Manufacturing*, 13 (2017) 1183–1190.
12. M. Javaid, A. Haleem, R. P. Singh, R. Suman, & S. Rab, Role of additive manufacturing applications towards environmental sustainability. *Advanced Industrial and Engineering Polymer Research*, 4 (2021) 312. https://doi.org/10.1016/j.aiepr.2021.07.005.
13. R. Godina, I. Ribeiro, F. Matos, B. T. Ferreira, H. Carvalho, & P. Peças, Article impact assessment of additive manufacturing on sustainable business models in industry 4.0 context. *Sustainability*, 12(17) (2020) 7066 (1–21).
14. H. Hegab, N. Khanna, N. Monib, & A. Salem, Design for sustainable additive manufacturing: A review. *Sustainable Materials and Technologies*, 35 (2023) e00576. https://doi.org/10.1016/j.susmat.2023.e00576.
15. B. Blakey-Milner, P. Gradl, G. Snedden, M. J. Brooks, J. Pitot, E. López, M. Leary, F. Berto, & A. du Plessis, Metal additive manufacturing in aerospace: A review. *Materials & Design*, 209 (2021) 110008. https://doi.org/10.1016/j.matdes.2021.110008.
16. M. Orme, I. Madera, M. Gschweitl, & M. Ferrari, Topology optimization for additive manufacturing as an enabler for light weight flight hardware. *Designs*, 2 (2018) 51. https://doi.org/10.3390/designs2040051.
17. G. Taddese, S. Durieux, & E. Duc, Sustainability performance indicators for additive manufacturing: A literature review based on product life cycle studies. *The International Journal of Advanced Manufacturing Technology*, 107 (2020) 3109. https://doi.org/10.1007/s00170-020-05249-2.
18. M. Kumaran, & V. Senthilkumar, Generative design and topology optimization of analysis and repair work of industrial robot arm manufactured using additive manufacturing technology. In *IOP Conference Series Materials Science and Engineering* (IOP Publishing, 2021), p. 12036. https://doi.org/10.1088/1757-899x/1012/1/012036.
19. A. W. Gebisa, & H. G. Lemu, A case study on topology optimized design for additive manufacturing. In *IOP Conference Series Materials Science and Engineering* (IOP Publishing, 2017), p. 12026. https://doi.org/10.1088/1757-899x/276/1/012026.
20. C. C. Tutum, S. Chockchowwat, E. Vouga, & R. Miikkulainen, Functional generative design. In H. Aguirre (Ed.), *Proceedings of the Genetic and Evolutionary Computation Conference* (Association for Computing Machinery, 2018), p. 1379. https://doi.org/10.1145/3205455.3205635.
21. N. Gardan, Knowledge management for topological optimization integration in additive manufacturing. *International Journal of Manufacturing Engineering*, 2014 (2014) 1. https://doi.org/10.1155/2014/356256.
22. A. M. Aliyi, & H. G. Lemu, Case study on topology optimized design for additive manufacturing. In *IOP Conference Series Materials Science and Engineering* (IOP Publishing, 2019), p. 12020. https://doi.org/10.1088/1757-899x/659/1/012020.
23. A. Balasubramanian, G. Shaktivel, & N. Raghukiran, Light-weighting of single-sided motorcycle swing-arm through generative design. In *IOP Conference Series Materials Science and Engineering* (IOP Publishing, 2021), p. 12016. https://doi.org/10.1088/1757-899x/1128/1/012016.
24. F. Laverne, R. Marquardt, F. Segonds, I. Koutiri, & N. Perry, Improving resources consumption of additive manufacturing use during early design stages: A case study.

International Journal of Sustainable Engineering, 12 (2019) 365. https://doi.org/10.1080/19397038.2019.1620897.

25. J. Coulthard, & C. Wang, Generative design and topology optimisation of products for additive manufacturing. In M. Shafik & K. Case (Eds.), *Advances in Transdisciplinary Engineering* (IOS Press, 2022). https://doi.org/10.3233/atde220570.
26. P. Nyamekye, R. Lakshmanan, V. Tepponen, & S. Westman, Sustainability aspects of additive manufacturing: Leveraging resource efficiency via product design optimization and laser powder bed fusion. *Heliyon*, 10 (2023) e23252. https://doi.org/10.1016/j.heliyon.2023.e23152.
27. B. H. Jared, M. A. Aguiló, L. L. Beghini, B. Boyce, B. W. Clark, A. Cook, B. Kaehr, & J. Robbins, Additive manufacturing: Toward holistic design. *Scripta Materialia*, 135 (2017) 141. https://doi.org/10.1016/j.scriptamat.2017.02.029.
28. M. Marek, & Z. Krejza, Sustainable building: Circular economy as a key factor for cost reduction. *E3S Web of Conferences*, 550 (2024) 1009. https://doi.org/10.1051/e3sconf/202455001009.
29. M. Italia, F. Papile, R. Santi, & B. D. Curto, Sustainable material selection framework: Taxonomy and systematisation of design approaches to sustainable material selection. *Sustainability*, 15 (2023) 16689. https://doi.org/10.3390/su152416689.
30. S. Chinchorkar, Adopting circular economy principles for electronic and photonic product recovery and recycling. In V. Katti, M. Pande, P. Jadhav, & A. Katti (Eds.), *Studies in Infrastructure and Control* (Springer Nature, 2025), p. 77. https://doi.org/10.1007/978-981-96-3065-3_5.
31. N. D. Ciceri, M. Garetti, & S. Sperandio, From product end-of-life sustainable considerations to design management. In B. Vallespir, & T. Alix (Eds.), *Advances in Production Management Systems. New Challenges, New Approaches. APMS 2009. IFIP Advances in Information and Communication Technology, Vol 338* (Springer Science+Business Media, 2010), p. 152. https://doi.org/10.1007/978-3-642-16358-6_20.
32. B. Rotondo, C. Bakker, R. Balkenende, & V. Arquilla, Integrating circular economy principles in the new product development process: A systematic literature review and classification of available circular design tools. *Sustainability*, 17 (2025) 4155. https://doi.org/10.3390/su17094155.
33. L. Sierra-Fontalvo, A. González-Quiroga, & J. A. Mesa, A deep dive into addressing obsolescence in product design: A review. *Heliyon*, 9 (2023) e21856 (1-15). https://doi.org/10.1016/j.heliyon.2023.e21856.
34. L. F. Dumée, Circular materials and circular design – review on challenges towards sustainable manufacturing and recycling. *Circular Economy and Sustainability*, 2 (2021) 9. https://doi.org/10.1007/s43615-021-00085-2.
35. J. A. Mesa, A. González-Quiroga, M. F. Aguiar, & D. Jugend, Linking product design and durability: A review and research agenda. *Heliyon*, 8 (2022) e10734 (1–15). https://doi.org/10.1016/j.heliyon.2022.e10734.
36. G. Mengesha, Beyond repair: A critical review of smart, sustainable, and AI-driven strengthening techniques for aging civil infrastructure. *SSRN Electronic Journal*, 13(6) (2025) 10–19. https://doi.org/10.2139/ssrn.5236159.
37. M. Gopal, H. G. Lemu, & E. M. Gutema, Sustainable additive manufacturing and environmental implications: Literature review. *Sustainability*, 15 (2022) 504. https://doi.org/10.3390/su15010504.
38. T. Peng, K. Kellens, R. Tang, C. Chen, & G. Chen, Sustainability of additive manufacturing: An overview on its energy demand and environmental impact. *Additive manufacturing*, 21 (2018) 694. https://doi.org/10.1016/j.addma.2018.04.022.

39. F. Psarommatis, G. May, & V. Azamfirei, Product reuse and repurpose in circular manufacturing: A critical review of key challenges, shortcomings and future directions. *Journal of Remanufacturing*, 15 (2025) 273–310. https://doi.org/10.1007/s13243-025-00153-y.
40. L. Kieush, J. Rieger, R. Attrotto, A. Sorino, W. V. der Stricht, H. Oterdoom, E. P. Heikkinen, G. Dall'Osto, C. Mapelli, D. Mombelli, L. D. Sante, F. Cirilli, V. Colla, T. A. Branca, I. Matino, A. Petrucciani, A. Zaccara, C. Brondi, E. Mousa, E. Nylund, E. Sandberg, M. Guzzon, E. Malfa, A. Schröder, & I. Bellemans, Roadmap for recycling practices and resource utilization in the iron and steelmaking industry: A case studies. *Matériaux & Techniques*, 112 (2024) 503. https://doi.org/10.1051/mattech/2024026.
41. L. G. L. M. Edirisinghe, A. A. P. de Alwis, & M. Wijayasundara, Sustainable circular practices in the textile product life cycle: A comprehensive approach to environmental impact mitigation. *Environmental Challenges*, 16 (2024) 100985. https://doi.org/10.1016/j.envc.2024.100985.
42. W. Wang, & M. M. Tseng, Economic analysis of product end-of-life strategies to achieve design for sustainable manufacturing. *5th International Conference on Responsive Manufacturing – Green Manufacturing (ICRM)* (2010) 268. https://doi.org/10.1049/cp.2010.0444.
43. I. Maslennikova, & D. Foley, Xerox's approach to sustainability. *INFORMS Journal on Applied Analytics*, 30 (2000) 226. https://doi.org/10.1287/inte.30.3.226.11666.
44. M. Kwak, & H. Kim, Market-driven positioning of remanufactured product for design for remanufacturing with part upgrade. *Proceedings of the ASME 2011 International Design Engineering Technical Conferences and Computers and Information in Engineering Conference*. Vol 9: 23rd International Conference on Design Theory and Methodology; 16th Design for Manufacturing and the Life Cycle Conference. Washington, DC, USA. August 28–31, 2011. ASME (2011) pp. 1043–1055. https://doi.org/10.1115/detc2011-48432.
45. F. M. A. Asif, M. Roci, M. Lieder, A. Rashid, A. Mihelič, & S. Kotnik, A methodological approach to design products for multiple lifecycles in the context of circular manufacturing systems. *Journal of Cleaner Production*, 296 (2021) 126534. https://doi.org/10.1016/j.jclepro.2021.126534.
46. J. A. C. Marfil, N. A. Serrat, E. L. Hitchen, & C. Viladecans-Riera, Recycling technology innovation as a source of competitive advantage: The sustainable and circular business model of a bicentennial company. *Sustainability*, 13 (2021) 7723. https://doi.org/10.3390/su13147723.
47. R. Aydın, & F. Badurdeen, Sustainable product line design considering a multi-lifecycle approach. *Resources Conservation and Recycling*, 149 (2019) 727. https://doi.org/10.1016/j.resconrec.2019.06.014.
48. P. Martínez, & I. Bosque, Sustainability dimensions: A source to enhance corporate reputation. *Corporate Reputation Review*, 17 (2014) 239. https://doi.org/10.1057/crr.2014.13.
49. Y. Li, & F.-S. Lin, Exploring design strategies for cultivating sustainability and enhancing brand image in personal care product brands. *Sustainability*, 16 (2024) 2476. https://doi.org/10.3390/su16062476.
50. M. R. Maarif, M. Syafrudin, & N. L. Fitriyani, Uncovering sustainability insights from amazon's eco-friendly product reviews for design optimization. *Sustainability*, 16 (2023) 172. https://doi.org/10.3390/su16010172.
51. W. Kong, A. Harun, R. S. Sulong, & J. Lily, The influence of consumers' perception of green products on green purchase intention. *International Journal of Asian Social Science*, 4 (2014) 924.

52. E. Loučanová, M. Šupín, T. Čorejová, K. Repková-Štofková, M. Šupínová, Z. Štofková, & M. Olšiaková, Sustainability and branding: An integrated perspective of eco-innovation and brand. *Sustainability*, 13 (2021) 732. https://doi.org/10.3390/su13020732.
53. V. Sahajwalla, & R. Hossain, Rethinking circular economy for electronics, energy storage, and solar photovoltaics with long product life cycles. *MRS Bulletin*, 48 (2023) 375. https://doi.org/10.1557/s43577-023-00519-2.
54. G. Hatcher, W. Ijomah, & J. F. C. Windmill, A network model to assist 'design for remanufacture' integration into the design process. *Journal of Cleaner Production*, 64 (2013) 244. https://doi.org/10.1016/j.jclepro.2013.09.021.
55. C. de Kwant, A. F. Rahi, & R. Laurenti, The role of product design in circular business models: An analysis of challenges and opportunities for electric vehicles and white goods. *Sustainable Production and Consumption*, 27 (2021) 1728. https://doi.org/10.1016/j.spc.2021.03.030.
56. J. A. Mesa, Design for circularity and durability: An integrated approach from DFX guidelines. *Research in Engineering Design*, 34 (2023) 443. https://doi.org/10.1007/s00163-023-00419-1.
57. E. Tyflopoulos, M. Lien, & M. Steinert, Optimization of brake calipers using topology optimization for additive manufacturing. *Applied Sciences*, 11 (2021) 1437. https://doi.org/10.3390/app11041437.
58. L. S. Sarma, C. Mallikarachchi, & S. Herath, Design-informed generative modelling using structural optimization. *arXiv (Cornell University)* (2023). https://doi.org/10.48550/arxiv.2308.06734.
59. W. Jiao, Y. Li, G. Hu, & M. Yang, Lightweight research in engineering: A review. *Applied Sciences*, 9 (2019) 5322. https://doi.org/10.3390/app9245322.
60. M. K. Niaki, S. A. Torabi, & F. Nonino, Why manufacturers adopt additive manufacturing technologies: The role of sustainability. *Journal of Cleaner Production*, 222 (2019) 381. https://doi.org/10.1016/j.jclepro.2019.03.019.
61. M. Jin, R. Tang, Y. Ji, F. Liu, L. Gao, & D. Huisingh, Impact of advanced manufacturing on sustainability: An overview of the special volume on advanced manufacturing for sustainability and low fossil carbon emissions. *Journal of Cleaner Production*, 161 (2017) 69. https://doi.org/10.1016/j.jclepro.2017.05.101.
62. J. L. Duigou, S. Gulbrandsen-Dahl, F. Vallet, R. Söderberg, B. Eynard, & N. Perry, Optimization and lifecycle engineering for design and manufacture of recycled aluminium parts. *CIRP Annals*, 65 (2016) 149. https://doi.org/10.1016/j.cirp.2016.04.111.
63. J. Joustra, B. Flipsen, & R. Balkenende, Circular design of composite products: A framework based on insights from literature and industry. *Sustainability*, 13 (2021) 7223. https://doi.org/10.3390/su13137223.
64. C. Spreafico, An analysis of design strategies for circular economy through life cycle assessment. *Environmental Monitoring and Assessment*, 194 (2022) 180. https://doi.org/10.1007/s10661-022-09803-1.
65. F. A. Halstenberg, K. Lindow, & R. Stark, Leveraging circular economy through a methodology for smart service systems engineering. *Sustainability*, 11 (2019) 3517. https://doi.org/10.3390/su11133517.
66. R. Kaur, R. Kumar, & H. Aggarwal, Systematic review of artificial intelligence, machine learning, and deep learning in machining operations: Advancements, challenges, and future directions. *Archives of Computational Methods in Engineering*, 32 (2025) 4983–5036. https://doi.org/10.1007/s11831-025-10290-z.
67. X. Zeng, A review on design of sustainable advanced materials by using artificial intelligence. *Deleted Journal*, 1 (2024) 10006. https://doi.org/10.35534/amsm.2024.10006.

68. M. Ghoreishi, & A. Happonen, Key enablers for deploying artificial intelligence for circular economy embracing sustainable product design: Three case studies. In *AIP Conference Proceedings* (AIP Publishing LLC, 2020), p. 050008. https://doi.org/10.1063/5.0001339.
69. M. Ghoreishi, & A. Happonen, New promises AI brings into circular economy accelerated product design: A review on supporting literature. *E3S Web of Conferences*, 158 (2020) 6002. https://doi.org/10.1051/e3sconf/202015806002.
70. F. Pedro, Design for sustainability and reconfigurable manufacturing systems–an critical analysis. *Journal of Machine and Computing*, 1 (2021) 11. https://doi.org/10.53759/7669/jmc202101002.
71. K. Karuppiah, B. Sankaranarayanan, & H.-W. Lo, A systematic literature review on the evolution of sustainable manufacturing practices: Key findings and implications. *Cleaner Engineering and Technology*, 22 (2024) 100798. https://doi.org/10.1016/j.clet.2024.100798.
72. R. R. Delgadillo, K. Medini, & T. Wuest, A DMAIC framework to improve quality and sustainability in additive manufacturing – A case study. *Sustainability*, 14 (2022) 581. https://doi.org/10.3390/su14010581.
73. S. Jung, L. B. Kara, Z. Nie, T. W. Simpson, & K. S. Whitefoot, Is additive manufacturing an environmentally and economically preferred alternative for mass production? *Environmental Science & Technology*, 57 (2023) 6373. https://doi.org/10.1021/acs.est.2c04927.

4 Additive Manufacturing Processes for Operational Sustainability

This chapter includes the importance of operational decisions in additive manufacturing (AM) and their influence on sustainability. It focuses on machine settings, build planning, energy consumption, material handling, avoidable waste on the shop floor and post-processing practices. Additionally, it also covers decisions around equipment, planning, recycling, post-processing, and supply chains. Additive manufacturing (AM) is a process of joining materials to create parts from 3D model data, usually layer by layer. AM includes standard process categories such as powder bed fusion, material extrusion, binder jetting, directed energy deposition, material jetting, sheet lamination, and vat photopolymerisation.

Operational sustainability involves more than just using 'green' materials. It also includes how a build is planned, the duration of machine operation, powder waste levels, and the amount of rework required after printing. Research shows that the impacts come from both the AM process and the steps involved in making and finishing the material. The chapter emphasises two main points. First, many additive manufacturing machines consume significant energy even when not printing. Second, many processes can reuse the same feedstock. However, reuse requires proper control to ensure parts remain consistent in quality [1].

4.1 ENERGY-EFFICIENT EQUIPMENT

Energy use in AM is strongly affected by the machine and its supporting systems. This includes heaters, chillers, inert gas systems, pumps, motion systems, and control electronics. ISO/ASTM 52900 also defines an 'additive system' as the AM machine plus its auxiliary equipment. These 'auxiliaries' often run continuously and account for much of the electricity use. In practice, auxiliaries can run for long periods, even when no material is being fused or deposited. This makes equipment choice a major sustainability decision. A detailed measurement study on polymer laser sintering shows that energy over time is the most important factor. In a detailed case study, a two-part build on a Sinterstation HiQ + HS took 1,308 minutes and used 88.68 kWh. During the build phase, the energy from time dependence was 53.95 kWh. The energy from the geometry dependence was 3.96 kWh. The energy from the Z-height dependence was 14.91 kWh [1]. The same parts on an EOSINT P 390 took 2301

DOI: 10.1201/9781003633754-4

minutes and used 103.16 kWh. This gap is not only about laser power. It is also about how each platform manages heating, cooling, and inert gas supply. The same study breaks the build into warm-up, build time, and cool-down. Warm-up energy was 8.69 kWh on the Sinterstation and 9.89 kWh on the EOSINT. Cool-down was 7.17 kWh on the Sinterstation and 16.92 kWh on the EOSINT. These are large 'fixed' loads. They occur even if only a few parts are in the build. This split explains why better insulation, tighter thermal control, and efficient cooling often matter more than small changes in scan settings.

Table 4.1 shows how energy is distributed across warm-up, build, and cool-down phases. A significant fraction of total energy demand is associated with non-productive phases, particularly warm-up and cool-down. These fixed energy loads become more important when machine utilisation is low or when build volumes are small. Warm-up durations differ substantially between the two systems. Although the EOSINT P 390 operates at a lower mean power during warm-up, its longer warm-up time leads to higher absolute energy consumption. Similar patterns are seen during the cooling phase, where longer cooling times significantly contribute to the total energy used by the job.

In both machines, the time it takes to build something is the main factor in how much energy is used. However, differences in the average power used during the build process and in how long it takes to build something affect overall machine efficiency. Longer build times mean that supporting systems, such as heaters, powder handling units, and control electronics, have to work for longer. As a result, the total energy needed for each task increases. The EOSINT P 390 consumes more energy per build than the Sinterstation HiQ + HS. The variance stems mainly from the EOSINT P 390's extended warm-up and cool-down periods.

This distinction is significant for equipment selection, as it explains why enhanced insulation, advanced temperature regulation, and efficient heating systems can

TABLE 4.1
Compares phase-wise energy consumption for two polymer laser sintering platforms: Sinterstation HiQ + HS and EOSINT P 390. The data are based on direct, instrumented energy measurements and are useful for both machine selection and production-level energy planning.

Build Phase	Metric	Sinterstation HiQ + HS	EOSINT P 390
Warm-Up	Duration (s)	6391	10266
	Energy (kWh)	8.69	9.89
	Mean power (kW)	4.89	3.47
Build Time	Duration (s)	60662	91783
	Energy (kWh)	72.82	76.35
	Mean power (kW)	4.32	2.99
Cool-Down	Duration (s)	11409	36007
	Energy (kWh)	7.17	16.92
Total Job	Total energy (kWh)	88.68	103.16

substantially reduce energy consumption compared with minor adjustments to laser power. Energy use per kilogram varies significantly across AM processes and machines. Reported values for electricity 'specific energy consumption' illustrate this range. For polymer processes, older yet widely cited measurements (from a later study) show that laser sintering consumes around 29.83–40.09 kWh/kg, stereolithography approximately 20.70–41.38 kWh/kg, and fused deposition modelling ranges from roughly 23.08 to 346.4 kWh/kg, depending on the system and conditions. For metal powder bed processes, reported values are about 31 kWh/kg for selective laser melting and about 17 kWh/kg for electron beam melting in comparative work [2].

Table 4.2 summarises representative specific energy consumption (SEC) ranges expressed in kWh/kg and highlights the key operational and process-related factors – such as machine warm-up, chamber heating, build height, laser or beam

TABLE 4.2
Comparison of typical specific energy consumption (SEC) values reported for major additive manufacturing (AM) technologies, classified according to ISO/ASTM process families.

AM Technology (Typical Name)	ISO/ASTM Process Family	Typical Reported SEC (kWh/kg)	What Drives the Value in Practice	Ref.
Laser Sintering (Polymer SLS/LS)	Powder bed fusion	~29.83–40.09 (≈14.5 also reported)	Long heat-up time, chamber heating, and build time dominate. Z-height and machine utilisation strongly affect SEC.	[1, 2]
Stereolithography (SLA)	Vat photopolymerisation	~20.70–41.38	Light source efficiency, curing strategy, and post-curing requirements influence energy use.	[2, 3]
Fused Deposition Modelling (FDM/FFF)	Material extrusion	~23.08–346.4	Machine warm-up, heated bed, and chamber power dominate, especially for small parts.	[2, 4]
Selective Laser Melting (SLM)	Powder bed fusion	~31	High laser power, inert gas circulation, and long processing time increase energy demand.	[3, 5]
Electron Beam Melting (EBM)	Powder bed fusion	~17	Efficient beam use and build strategy reduce SEC, though build height still matters.	[3, 5]

characteristics, and machine utilisation – that influence energy consumption in practice. Reported values are compiled from representative studies in the literature and may vary depending on machine configuration, material system, part geometry, and build strategy.

The sources include reported ranges for polymer processes such as selective/laser sintering (SLA/LS), stereolithography (SLA), and fused deposition modelling/fused filament fabrication (FDM), as well as compiled SEC values for various additive manufacturing methods such as selective laser melting (SLM) and electron beam melting (EBM).

Practical steps to enhance energy efficiency are often straightforward to implement. It is needed to select equipment sizes based on typical part dimensions to prevent the build volume from remaining largely empty during operation. Accessories, including filters, optics, and recoater systems, must be meticulously maintained to avoid failed builds that trigger prolonged and unnecessary heating cycles. Furthermore, it uses power metres on each individual machine rather than only at the plant level to accurately identify the subsystems with the highest energy consumption. This case study evidence underscores the importance of this approach. It shows that the highest energy usage often comes from systems that operate independently of geometry-scanning activities [1].

4.2 PROCESS OPTIMISATION

It discusses how to run the process to reduce energy use, scrap, and rework. In additive manufacturing, poor planning can turn a process that produces little waste into one that produces a lot. This can happen due to failed builds, too many support requests, and long post processing times. A key concept is capacity utilisation. If a machine spends a lot of time heating but only produces a few parts, the energy used per part becomes very high. In laser sintering, it is important to evaluate energy efficiency by considering how much of the machine's capacity is used and the build height. These factors significantly influence the amount of energy required. The laser sintering case study provides a useful way to explain batching. Most build-phase energy was time-dependent (53.95 kWh) and Z-height-dependent (14.91 kWh). Geometry-dependent energy was small (3.96 kWh) for the two-part build. Warm-up and cool-down energy were also fixed per job [1].

Below is a straightforward illustration (Figure 4.1) of batching by 'same height'. It assumes additional parts are added within the same build height, maintaining a consistent Z-height. The measured energy distribution is used to demonstrate how fixed energy is divided among more parts. This serves as an operational example rather than a universal principle.

Therefore, the practical application of this process is straightforward. If parts can be nested without increasing the height of the tallest part, then the energy used per part can decrease. This aligns with the observation that building space utilisation is critical for accurate energy comparisons in AM. Process settings also matter. A review of manufacturing energy notes that energy process settings are important. Energy-saving strategies in manufacturing include reducing the amount of support material used in material extrusion and lowering the height of parts made with laser sintering. The cited research shows that these methods can save up to 61% of energy

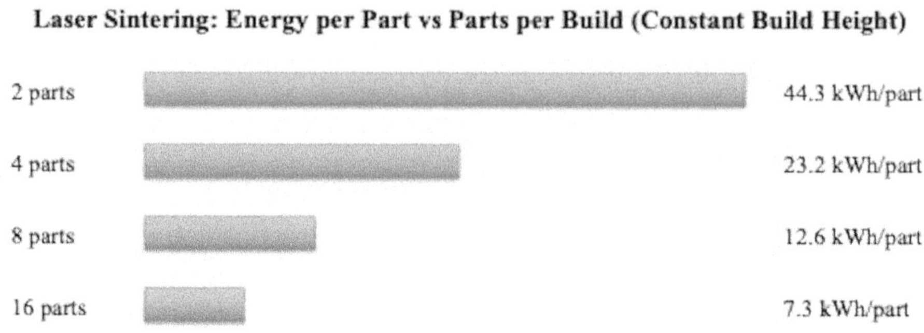

FIGURE 4.1 Illustrative relationship between energy consumption per part and the number of parts produced in a single laser sintering (LS) build at a constant build height. As the number of parts per build increases, the energy consumption per part decreases non-linearly due to the distribution of fixed machine energy, such as warm-up, chamber heating, and standby losses, across a larger number of components. This trend highlights the strong influence of build utilisation, packing density, and build height on the energy efficiency of powder bed fusion processes. Similar utilisation-driven effects have been widely reported for laser sintering systems in [1, 3, 5].

in fused deposition modelling (FDM) and 43% in selective laser sintering (SLS). FDM provides a straightforward example of this. In one measured case, printing one standard sample consumed 717.1 Wh, and warm-up alone consumed 435.5 Wh. The same source notes that if multiple parts are fabricated after warm-up, SEC drops [2]. This supports a common shop-floor rule: batch small FDM jobs when possible and avoid repeated cold starts. Optimisation needs balance. Design changes like lattices can reduce mass, but they can also increase build time and energy if they increase scan length or require more supports. A life cycle assessment of lattice structures reports that lattice designs reduce weight but can increase 'material usage, energy consumption, and environmental impact' due to higher volume and processing requirements in some cases [3].

4.3 RECYCLING AND MATERIAL REUSE

It addresses the recycling of powders, resins, and polymers, and discusses methods for managing reuse to maintain process stability. Powder-based AM has a clear advantage. Much of the feedstock is not fused in each build. In polymer powder bed fusion, unused powder supports the part and can often be collected and reused. In metal powder bed fusion, unused powder can also be recovered through sieving and controlled handling [3].

Reuse is not automatic; powders degrade over time due to heat exposure, oxygen absorption, and moisture uptake. A review on electron beam powder bed fusion highlights increasing oxygen levels with repeated reuse, citing an example where oxygen content rose from 0.08 wt.% to 0.19 wt.% after 21 reuse cycles. This is a quality risk that can also become a sustainability issue because quality drift leads to scrap or conservative 'throw-away' practices. Official guidance from printer vendors on materials helps quantify reuse through refresh rate, which represents the proportion of virgin powder added back to maintain stable properties. For polymer SLS, a vendor material page for Nylon 12 powder lists refresh rates such as 30% (and other values depending on the powder grade). A 30% refresh rate implies up to 70% of the powder mix can come from recovered powder in that cycle, subject to screening and conditioning.

For Multi Jet Fusion, the manufacturer's material documentation highlights high reuse ratios. HP documentation describes powder reusability and notes reuse ratios that can reach high values (commonly cited around 80% for 'high reusability' PA12 setups), with the broader claim that some materials support reuse ranges of 60–90%, depending on material and workflow [6]. Reuse is also linked to waste reduction. An open-access academic paper on recycling SLS-waste nylon highlights a prevalent industry challenge: it reports a typical SLS build volume of approximately 20 kg and estimates that around 25% of the powder, roughly 5 kg, is waste generated per build [7]. Even if the exact percentage varies between machines and powders, the key lesson remains the same: improved refresh rules, better mixing, and more effective powder handling directly reduce waste.

$$E_{part} = \frac{E_{fixed} + E_{process}}{N}$$

where:

E_{part} = energy consumed per part (kWh/part)
E_{fixed} = fixed machine energy (warm-up, chamber heating, standby)
$E_{process}$ = laser scanning and recoating energy
N = number of parts per build

For a constant build height, fixed thermal and machine-related energy dominate the total energy demand. Increasing the number of parts per build distributes the fixed energy across more components, resulting in a significant reduction in energy consumption per part.

Table 4.3 summarises typical reuse or refresh ratios reported in the literature and highlights the key process and material control parameters required to maintain part quality during repeated feedstock reuse. Reported practices assume good housekeeping, including effective sieving or filtration, controlled storage, and systematic tracking of material history. Reuse limits and quality sensitivity vary with feedstock type, process conditions, and machine configuration.

4.4 POST-PROCESSING EFFICIENCY

It deals with the steps after printing. Common steps include de-powdering, support removal, washing, curing, blasting, machining, heat treatment, and inspection. Here, note that two points matter for sustainability. First, post-processing can be

TABLE 4.3
Comparison of material reuse and refresh practices across additive manufacturing (AM) processes, classified according to ISO/ASTM process categories.

ISO/ASTM Process Category	Feedstock Type	Typical Reuse/ Refresh Practice	Quality-Critical Control Parameters	References
Powder Bed Fusion – Polymers (PBF-P)	PA12 and similar polymer powders	Refresh ratios typically range from 20% to 50%. For Nylon-12, ~30% fresh powder addition is common, allowing ~70% reused powder.	Moisture control, sieving efficiency, thermal history tracking, consistent powder mixing	[1, 2]
Powder Bed Fusion – Polymers (PBF-P, Multi Jet Fusion)	Polymer powders	High reuse ratios of ~60–90% are reported for suitable materials under controlled conditions.	Powder conditioning, contamination prevention, reuse tracking, material ageing control	[3]
Powder Bed Fusion – Metals (PBF-LB/ PBF-EB)	Metal powders	High physical powder recovery is possible, but chemical and morphological changes limit reuse cycles.	Oxygen/nitrogen pickup, particle size drift, spatter removal, controlled storage	[4, 5]
Material Extrusion (MEX)	Thermoplastic filament	Scrap and failed prints can be mechanically reprocessed into filament in limited setups.	Polymer degradation, filament diameter consistency, and contamination control	[8]
Vat Photopolymerisation (VPP)	Liquid photopolymer resin	Resin reuse is possible with filtration, but ageing and exposure limit long-term reuse.	Resin ageing, viscosity drift, contamination, and cure inhibition	[9]

essential, so it must be made efficient, not ignored. Second, post-processing often benefits from batching, as does printing. The ISO/ASTM vocabulary notes that the removal of support structures and cleaning are often necessary, though they may not be counted as separate process steps in some definitions of single-step versus multi-step AM processes [10]. In real factories, these steps consume labour, electricity, and consumables. Peer-reviewed work on AM environmental impact explicitly includes post-treatment in discussions of the overall impact of AM manufacturing chains [11]. Post-processing reviews also emphasise that these steps are used to improve surface finish and mechanical performance, especially for metallic AM parts [12]. Many post-processing steps are not reported as clearly as printing energy. Still, practical ranges exist for key steps, especially for thermal treatments.

Table 4.4 gives the indicative electricity per kg treated for selected steps. Values are derived from studies on manufacturing and heat-treatment energy, as well as from lifecycle assessment work modelling HIP [13].

Compared with the specific energy consumption (SEC) of additive manufacturing processes, post-processing steps represent a significant and sometimes comparable contribution to overall energy demand. Reported printing-stage SEC values for common AM processes typically range from tens of kilowatt-hours per kilogram, depending on machine utilisation and build strategy. In comparison, individual post-processing operations such as stress relief, hardening, carburising, or hot isostatic pressing add several additional kilowatt-hours per kilogram. While these values are often lower than the printing-stage SEC, poor furnace loading, repeated thermal cycles, or mandatory densification treatments can substantially increase the total energy footprint of an AM component. Accordingly, sustainability assessments of AM should consider both printing and post processing stages together, as design and process decisions that minimise post-processing requirements can lead to meaningful reductions in overall energy consumption.

Table 4.4 shows indicative electricity consumption values, expressed in kWh per kilogram of material treated, for common post-processing steps used in additive manufacturing. These steps include thermal treatments, densification, and finishing operations that are often applied to improve the dimensional stability, mechanical performance, and surface quality of AM components. The reported values are based on manufacturing energy studies and life cycle assessment (LCA) literature and reflect typical operational ranges rather than exact figures. Actual energy use may vary depending on equipment type, batch size, furnace utilisation, cycle design, and material specification. The table also highlights key operational controls that affect energy efficiency and material quality during post-processing.

Practical measures that improve post-processing efficiency are mostly workflow measures. Design to reduce supports, because support removal is wasted material and time. Combine similar alloys and sizes so that heat treatment and HIP runs are fully utilised. Use robust de-powdering and sieving workflows to ensure high powder recovery and safety. The main point is supported by an LCA framing: impacts stem from both the AM unit process and the post-treatment chain.

TABLE 4.4
Indicative electricity demand for selected post-processing steps in additive manufacturing.

Post-Processing Step	Purpose in AM	Indicative Energy (kWh/kg)	Operational Control Notes	Refs.
Stress Relief/ Direct Hardening (Protective Atmosphere)	Reduce residual stresses; stabilise dimensions	~1.01	Maintain high furnace utilisation; avoid partially loaded cycles	[5]
Vacuum Hardening	Improve hardness and microstructural uniformity	~1.93	Optimise heat-treatment cycles; minimise repeated runs	[5]
Carburising	Surface hardening of steel components	~2.54	Energy-intensive; apply only when required by specification	[5]
Hot Isostatic Pressing (HIP)	Reduce internal porosity; improve fatigue performance	~3.95 (≈14.21 MJ/kg)	Batch components to increase load factor; justify use by performance gains	[3]
CNC Finish Machining (Example Case)	Achieve dimensional tolerance and surface finish	~3.49	Minimise material removal; apply near-net-shape AM design	[1]

4.5 ON-DEMAND MANUFACTURING AND LOCAL PRODUCTION

This section refers to production models, not just machines. AM can support on-demand manufacturing because a digital file can replace stored physical inventory for many parts. This can cut waste from obsolete stock. It can also reduce transport emissions by producing parts closer to their point of use. A widely cited life cycle assessment of distributed 3D printing compared distributed manufacturing with conventional manufacturing for polymer products. The study reports that cumulative energy demand could be reduced by about 41–64% in the assessed cases, with larger reductions when powered by photovoltaics [14]. This result varies depending on what is being compared, but it illustrates why local production is often considered a key factor for sustainability. Transport emissions are a major part of the 'local production' argument. The official UK government methodology provides air freight emission factors per tonne-km. For 2025 factors, 'all air freight' is reported at about 0.49 kg CO_2e per tonne-km for long-haul, about 0.70 for short-haul, and about 2.52 for domestic, on the stated basis [15] (Figure 4.2).

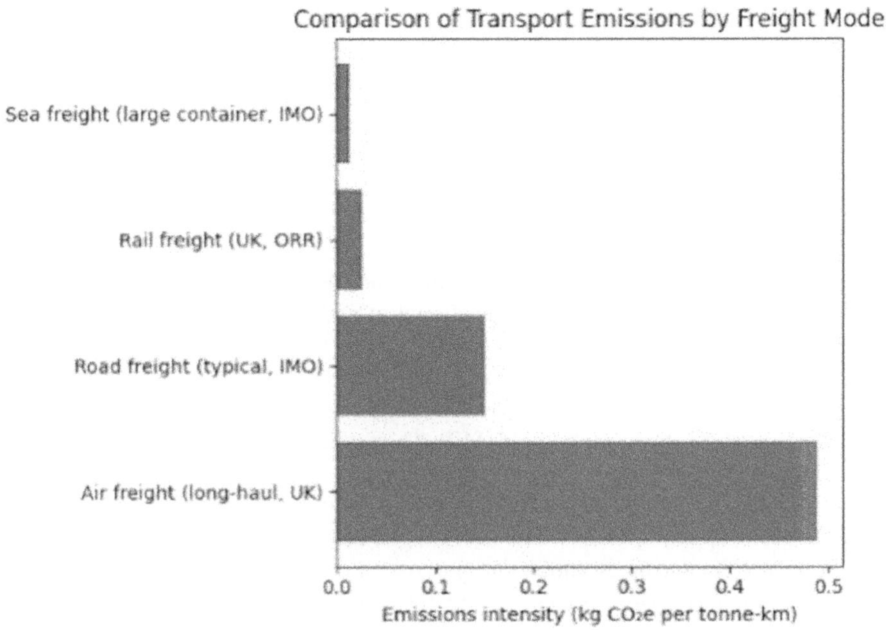

FIGURE 4.2 Indicative CO_2 emissions intensity of major freight transport modes. Air freight shows substantially higher emissions per tonne-kilometre compared with road, rail, and sea transport, explaining why local production strategies that avoid air freight can deliver significant carbon savings. Indicative emissions intensity (kg CO_2e per tonne-km) One bar ≈ 0.05 kg CO_2e/tkm.

The International Maritime Organization (IMO) offers a look at carbon intensity across various transportation methods, measured in grams of CO_2 per tonne-kilometre. Road freight generally ranges from 80 to 180 g CO_2/tkm, with the average landing at around 150 g CO_2/tkm. In contrast, large container ships, those with a capacity exceeding 8,000 TEU, demonstrate significantly better transport efficiency. Their reported emission intensities are roughly 12.5 g CO_2/tkm, while smaller vessels show somewhat higher values [16]. This contrast can be illustrated with a simple example. Shipping a 1 kg spare part across 8,000 km via long-haul air freight translates to 8 tonne-kilometres of transport.

Using an emission factor of 0.49 kg CO_2e/tkm, the resulting transport emissions are approximately 3.9 kg CO_2e, even before accounting for packaging, handling, and warehousing [16]. By comparison, transporting the same part by a large container ship at approximately 0.0125 kg CO_2e/tkm results in transport emissions of only about 0.10 kg CO_2e. Even with the inclusion of additional road routes in real-world logistics networks, the scale difference between air and sea freight remains significant. In contrast, producing goods locally doesn't automatically guarantee a smaller environmental impact [16].

The advantages can be significantly diminished, or even entirely offset, by factors such as inefficient equipment use, increased material waste, or reliance on electricity generated by carbon-intensive sources. Therefore, the most substantial sustainability improvements in on-demand additive manufacturing are realised when it is integrated with high build utilisation, rigorous process control, and comprehensive material recovery and recycling protocols.

4.6 BEYOND 3D PRINTING – EMERGING TECHNOLOGIES FOR SUSTAINABILITY

The section looks beyond current 3D printing practice. The main idea is that sustainability can also come from smarter functionality and fewer process steps, not only from lower kWh.

4D printing is one emerging direction. The core idea is that printed structures can change shape or function over time in response to external stimuli such as temperature, light, moisture, or magnetic fields [17]. The sustainability aspect is practical. If a component can self-deploy, self-adjust, or self-seal, it can reduce fasteners, lower part counts, and decrease maintenance actions. These are indirect benefits, but they can be significant for field equipment where servicing is costly [18].

Hybrid manufacturing is another important path. It combines additive and subtractive manufacturing on a single platform. It can deposit material and then machine it in the same setup. This reduces re-clamping, handling, and scrap caused by misalignment between steps. Reviews argue that hybrid platforms can streamline workflows while keeping the design freedom of AM and the accuracy of Computer Numerical Control machining [19]. For sustainability, the direct benefit is better 'right-first-time' quality, because fewer handoffs usually mean fewer errors and less rework.

In-situ monitoring and closed-loop control are also important. The role in sustainability is straightforward. Detecting defects early reduces failed builds, saving powder, time, and energy. Recent research explicitly links in-situ monitoring and closed-loop control to improved reliability, reducing failure rates and the wasted energy associated with them [20]. In operational terms, these emerging technologies still rely on the same fundamentals in this chapter. They need efficient equipment, good planning, controlled reuse, and efficient post-processing. The difference is that they offer new ways to reduce waste at the system level, such as fewer assemblies, fewer repairs, and fewer logistics movements.

REFERENCES

1. M. Baumers, C. Tuck, D. L. Bourell, R. Sreenivasan, & R. Hague, Sustainability of additive manufacturing: Measuring the energy consumption of the laser sintering process. *Proceedings of the Institution of Mechanical Engineers, Part B: Journal of Engineering Manufacture*, 225(12) (2011) 2228–2239. https://doi.org/10.1177/0954405411406044.
2. H.-S. Yoon, J.-Y. Lee, H.-S. Kim, M.-S. Kim, E.-S. Kim, Y.-J.Shin, W.-S. Chu, & S.-H. Ahn, A comparison of energy consumption in bulk forming, subtractive, and additive processes: Review and case study. *International Journal of Precision Engineering and Manufacturing–Green Technology*, 1(3) (2014) 261–279. https://doi.org/10.1007/s40684-014-0033-0.

3. K. Kellens, M. Baumers, T. G. Gutowski, W. Flanagan, R. Lifset, & J. R. Duflou, Environmental dimensions of additive manufacturing: Mapping application domains and process characteristics. *Journal of Industrial Ecology*, 21(S1) (2017) S49–S68. https://doi.org/10.1111/jiec.12629.
4. T. Peng, K. Kellens, R. Tang, C. Chen, & J. R. Duflou, Sustainability of additive manufacturing: An overview on its energy demand and environmental impact. *Additive Manufacturing*, 21 (2018) 694–704. https://doi.org/10.1016/j.addma.2018.03.003.
5. M. Baumers, P. Dickens, C. Tuck, & R. Hague, The cost of additive manufacturing: Machine productivity, economies of scale, and technology push. *Technological Forecasting and Social Change*, 102 (2016) 193–201. https://doi.org/10.1016/j.techfore.2015.02.015.
6. HP Development Company, L.P. (n.d.). HP 3D printing materials (Brochure No. 4AA7-1533ENE). HP. Retrieved February 6, 2026, from https://h20195.www2.hp.com/V2/getpdf.aspx/4AA7-1533ENE.pdf.
7. M. Uddin, D. Williams, & A. Blencowe, Recycling of selective laser sintering waste nylon powders into fused filament fabrication parts reinforced with Mg particles. *Polymers*, 13(13) (2021) 2046. https://doi.org/10.3390/polym13132046.
8. M. Kreiger, & J. M. Pearce, Environmental life cycle analysis of distributed three-dimensional printing and conventional manufacturing of polymer products. *Journal of Cleaner Production*, 61 (2013) 224–236. https://doi.org/10.1016/j.jclepro.2013.06.016.
9. T. D. Ngo, A. Kashani, G. Imbalzano, K. T. Q. Nguyen, & D. Hui, Additive manufacturing (3D printing): A review of materials, methods, applications and challenges. *Composites Part B: Engineering*, 143 (2018) 172–196. https://doi.org/10.1016/j.compositesb.2018.02.012.
10. ISO/ASTM 52900:2021. (2021). Standard terminology for additive manufacturing – General principles – Terminology (ISO/ASTM 52900:2021). ISO & ASTM International. www.iso.org/obp/ui/en/#iso:std:74514:en.
11. K. Kellens, R. Mertens, D. Paraskevas, W. Dewulf, & J. R. Duflou, Environmental impact of additive manufacturing processes: Does AM contribute to a more sustainable way of part manufacturing? *Procedia CIRP*, 61 (2017) 582–587. https://doi.org/10.1016/j.procir.2016.11.153.
12. J. Su, W. L. Ng, J. An, W. Y. Yeong, C. K. Chua, & S. L. Sing, Achieving sustain ability by additive manufacturing: A state-of-the-art review and perspectives. *Virtual and Physical Prototyping*, 19(1) (2024) e2438899. https://doi.org/10.1080/17452759.2024.2438899.
13. C. H. Mbanyeudei. (2023). Heat treatment energy mapping (Master's thesis). Swedish University of Agricultural Sciences, DiVA Open Access Repository.
14. M. Kreiger, & J. M. Pearce, Environmental life cycle analysis of distributed three-dimensional printing and conventional manufacturing of polymer products. *ACS Sustainable Chemistry & Engineering*, 1(12) (2013) 1511–1519. https://doi.org/10.1021/sc400093k.
15. Department for Environment, Food & Rural Affairs. (2025). Greenhouse gas reporting: Conversion factors methodology paper 2025 (Version 1.0). UK Government. https://assets.publishing.service.gov.uk/media/6846b0870392ed9b784c0187/2025-GHG-CF-methodology-paper.pdf.
16. International Maritime Organization. (2009). Second IMO greenhouse gas study 2009 (MEPC 59/INF.10). IMO. https://wwwcdn.imo.org/localresources/en/OurWork/Environment/Documents/MEPC%2059-INF.10%20-%20Second%20IMO%20GHG%20Study%202009.pdf.
17. P. E. Antezana, S. Municoy, G. O. Ostapchuk, P. N. Catalano, J. G. Hardy, P. A. Evelson, G. Orive, & M. F. Desimone, 4D printing: The development of responsive

materials using 3D-printing technology. *Pharmaceutics*, 15(12) (2023) 2743. https://doi.org/10.3390/pharmaceutics15122743.
18. R. Sajjad, S. T. Chauhdary, M. T. Anwar, A. Zahid, A. A. Khosa, M. Imran, & M. H. Sajjad, A review of 4D printing – Technologies, shape shifting, smart polymer based materials, and biomedical applications. *Advanced Industrial and Engineering Polymer Research*, 7(1) (2024) 20–36. https://doi.org/10.1016/j.aiepr.2023.08.002.
19. B. Freitas, V. Richhariya, M. Silva, A. Vaz, S. F. Lopes, & Ó. Carvalho, A review of hybrid manufacturing: Integrating subtractive and additive manufacturing. *Materials*, 18(18) (2025) 4249. https://doi.org/10.3390/ma18184249.
20. J. Su, W. L. Ng, J. An, W. Y. Yeong, C. K. Chua, & S. L. Sing, (2024). Achieving sustainability by additive manufacturing: A state-of-the-art review and perspectives. *Virtual and Physical Prototyping*, 19(1), e2438899. https://doi.org/10.1080/17452759.2024.2438899.

5 Novel Approaches and Applications for Operational Sustainability

Novel and developing strategies are being explored to improve the sustainability of manufacturing processes and the products they create. These strategies include using renewable and recycled materials, choosing polymers that can break down or compost in suitable conditions, reducing the need for new metal extraction through better powder management, and improving product quality while minimising waste through better monitoring and control during the printing process.

These strategies shift the focus towards solving global environmental problems, such as climate change and resource depletion. These approaches also improve the sustainability of manufacturing, especially in additive manufacturing. The focus on sustainability reduces the environmental impact of industrial activities. These approaches minimise waste, save energy, lessen environmental burdens and transportation costs, and extend the lifespan of materials by using options such as bio-based or recycled materials, continuous fibre composites, and local recycling. Additive manufacturing redefines traditional manufacturing. It allows precise and timely production. Further, it enables user-friendly, efficient, durable, and recyclable designs using new eco friendly approaches. With rapid prototyping and on-demand production capabilities, these innovative approaches reduce inventory costs and energy use in warehousing and logistics. In essence, these novel approaches will enable manufacturing to move towards a more sustainable future by minimising environmental harm and maximising resource efficiency.

Additive manufacturing (AM) creates parts by adding material layer by layer, following the information in a digital model. This is different from machining, where a large block is cut down into the final shape. This basic shift changes how waste, energy use, and material choice behave across the full process chain. Operational sustainability in AM means three simple things. First, use less material and generate less scrap. Second, reduce energy per 'good part' by reducing print failures and rework. Third, avoid harmful materials and unsafe handling, especially for powders and fumes. These points are repeated across AM sustainability reviews and standards work. The literature states that AM is not 'automatically green'. AM can reduce material waste and enable lighter designs, but the specific energy consumption (SEC) can be high, especially when machine utilisation is low or when many parts fail and need reprinting. Measured and reviewed SEC ranges vary widely across process

DOI: 10.1201/9781003633754-5

categories, materials, machines, and operating strategies. The core difference that drives many sustainability outcomes: subtractive manufacturing removes material, while AM places material only where needed and can also create internal lattices and part-consolidation features.

The manufacturing industries need to change their standard practices to be more eco-friendly due to rising environmental challenges and the availability of limited non-renewable resources. This change is essential for maintaining economic stability and social health. To reach operational sustainability, companies must adopt a comprehensive approach that integrates environmentally friendly design principles at every stage of a product's life, from initial concept to disposal. It is important to consider economic, ecological, and social factors together, moving beyond traditional manufacturing methods. This shift involves rethinking current production approaches and emphasising sustainable practices and efficient use of resources. Sustainable manufacturing is driven by Industry 5.0, which integrates digital transformation and innovative business models to improve corporate sustainability.

This chapter focuses on seven practical innovation areas that directly affect operational sustainability: (i) sustainable polymer feedstocks, (ii) biodegradable and compostable polymers, (iii) sustainable metal powders, (iv) continuous fibre composites, (v) distributed recycling and circular supply, (vi) hybrid and multi-material printing, and (vii) in situ monitoring and process control.

5.1 BIO-BASED AND RECYCLED MATERIALS

Bio-based polymers are attractive because they can reduce dependence on fossil feedstocks and may reduce environmental impacts. Bio-based constituents are also being explored in AM materials. They include bio-based fillers and bio-derived resin components. In AM, the most common bio-based polymer is polylactic acid (PLA). PLA is popular for material extrusion (also called fused filament fabrication (FFF)) because it is easy to process and inexpensive. Many studies also explore PLA biocomposites that use plant-based fillers such as cellulose, lignin, and other biomass residues. These fillers can reduce the use of virgin polymer and improve stiffness, but they can also increase brittleness and moisture sensitivity [1].

For operational sustainability, recycled polymers matter as much as bio-based polymers. Recycled polylactic acid (PLA), polyethene terephthalate glycol (PET/PETG), acrylonitrile butadiene styrene (ABS), PP, and high-density polyethylene (HDPE) have all been used to produce filaments or pellets for extrusion-based AM. The main advantage is straightforward: recycled polymers usually require less energy and produce fewer emissions than producing virgin resins, especially when transport distances are reduced, and recycling systems are efficient [2].

Recycled materials directly support circular economy goals by keeping materials in use and reducing the need for virgin materials. A circular economy approach is commonly summarised as: reduce waste and pollution, keep products/materials in circulation, and regenerate nature. Recycled plastics often vary in quality because of contamination, mixed polymers, moisture, and damage from repeated heating.

These factors can change strength, layer bonding, and print reliability. Bio-based and biodegradable plastics may also contain additives (for processing, colour, flexibility, and stability). Additives matter for safety, performance, and end-of-life outcomes, including possible chemical migration in some uses.

At the same time, recycled polymer feedstocks are harder to control. Repeated melting and extrusion can reduce molecular weight and change viscosity. This can alter layer bonding and increase the risk of brittle failure, warping, or poor surface finish. Reviews of recycled filament materials repeatedly highlight that sorting, cleaning, drying, and process control are the key factors that differentiate stable from unstable printing [2].

A practical example is large-format extrusion using recycled granules. Studies that track multiple recycling cycles of PLA show measurable degradation trends but also show that careful process design can keep performance within useful ranges for many applications. This is important for furniture, display parts, jigs, and non-safety-critical structures, where the goal is to keep material loops local and reduce demand for virgin feedstock [3].

Bio-based and recycled materials signify a pivotal transformation in the procurement of materials for manufacturing, especially within the realm of additive manufacturing. These materials present feasible alternatives to traditional plastics derived from petroleum, encompassing ABS and PETG, both of which are associated with a significant carbon footprint and inadequate end-of-life disposal methods. The use of bio-based materials, like those made from cellulose or chitin, along with recycled polymers, directly addresses the growing problem of plastic waste. At the same time, this approach supports a circular economy in additive manufacturing [4, 5].

Therefore, this method reduces the need to extract new fossil fuels and lessens the environmental problems related to disposing of traditional polymers. This is consistent with the principles of industrial ecology and resource efficiency [6]. The growing use of additive manufacturing in many industries highlights the need for sustainable materials. This is supported by a recent survey that showed a significant increase in the use of 3D printing for production purposes [7].

Bio-based materials, like PLA, come from renewable resources such as corn or sugarcane, or from microbial sources, as seen with polyhydroxyalkanoates (PHA). Conversely, recycled materials originate from waste generated by both consumers and industrial sectors. This encompasses recycled polyethene terephthalate (PET) derived from plastic containers and recycled acrylonitrile butadiene styrene (ABS) sourced from electronic refuse. Studies suggest that using recycled PET (rPET) can significantly reduce greenhouse gas emissions. This reduction could be as much as 60% compared to making new PETG. This example highlights a significant environmental benefit, indicating a substantial decrease in emissions. Furthermore, advances in effective recycling methods for additive manufacturing waste, including classifying recycled powder by melt flow rate and evaluating reuse cycles, enhance material circularity and reduce dependence on raw materials [8].

This is of paramount importance given the constraints on the availability of recycled AM materials, despite the sector's rapid progress towards more sustainable feedstocks, such as biopolymers and bio-based plastics derived from natural materials like algae or orange peels [9]. These developments directly address crucial environmental issues linked to conventional manufacturing by providing avenues to minimise waste and lessen reliance on limited resources, thereby adhering to circular economy principles [6, 10].

Polylactic Acid, a bioplastic, is made from corn or sugarcane. This renewable material breaks down naturally and finds applications in packaging, prototyping, and medical devices. Microorganisms synthesise bioplastics like polyhydroxyalkanoates, which are valued for their renewable and partially biodegradable properties, making them ideal for packaging, prototypes, and medical applications. Recycled PET (rPET) is derived from used items such as bottles and transformed into filament. Discarded electronics serve as raw material for recycled acrylonitrile butadiene styrene (ABS), which is then converted into filament. Incorporating natural fibres like hemp, flax, or jute with plastics such as PLA can enhance durability and lessen environmental impact.

Table 5.1 compares bio-based and petroleum-based polymers commonly used in additive manufacturing with respect to their origin, end-of-life behaviour, mechanical performance, thermal resistance, carbon footprint, and typical applications. Bio-based polymers are primarily derived from renewable resources and generally exhibit lower environmental impact, although they often possess moderate strength and lower heat resistance. In contrast, petroleum-based polymers are sourced from fossil fuels and offer higher mechanical strength and thermal stability but are associated with higher carbon emissions and limited end-of-life options. The comparison highlights the trade-offs involved in material selection for sustainable additive manufacturing and underscores the growing interest in bio-based alternatives for environmentally responsible applications [1, 2, 8].

TABLE 5.1
Comparison of bio-based polymers and petroleum-based polymers used in additive manufacturing.

Feature	Bio-Based Polymers	Petroleum-Based Polymers
Origin	Renewable crops (e.g. corn, sugarcane, microbes)	Fossil fuels
End-of-Life	Industrially compostable or partly biodegradable	Not biodegradable; high carbon footprint
Strength	Moderate strength	Higher strength
Heat Resistance	Lower heat resistance (e.g. PLA softens at low temperatures)	Higher thermal stability
Carbon Footprint	Lower	Higher
Typical Uses	Prototyping, medical devices, packaging	Tooling, durable parts, fixtures (ABS, PETG)

5.2 BIODEGRADABLE AND COMPOSTABLE POLYMERS

Biodegradable and compostable polymers are widely discussed as 'greener plastics', but the real sustainability outcome depends on where and how they break down. In additive manufacturing (AM), these materials are mostly used in material extrusion (FFF/fused deposition modelling (FDM)) because filaments and pellets are easy to handle and low-cost. The final performance of a part is affected by the polymer chemistry used, the printing settings, and the part's design, including factors such as wall thickness, fill pattern, and surface area. 'Biodegradable' material can be broken down by microorganisms, but the rate and completeness depend on the environment (temperature, oxygen, moisture, microbes, and time).

"Compostable" is a more technically precise term, associated with standardized testing protocols that quantify compostability beyond mere biodegradation.. They require evidence (via standard tests) of biodegradation and disintegration during composting, and the absence of harmful effects on the composting process and on compost quality [11]. In Europe, EN 13432 is widely used for industrial compostability claims for packaging. It also uses a test scheme that checks biodegradation, disintegration, and compost quality [12].

Testing compostability often uses controlled thermophilic conditions, which is a common method in industrial composting. ISO 14855 is a core method for determining 'ultimate aerobic biodegradability' under controlled composting by measuring the amount of evolved CO_2. ISO 20200 is a related method, but it measures only physical disintegration (breakdown). It is not enough to claim compostability; ISO explicitly states that further testing is required for such claims. For real-world decision-making, the key point is simple: a standard test result is conditional. It is valid for the tested material and form under the defined test conditions, and performance can change with part thickness, additives, and printing method.

Industrial Composting vs. Home Composting: Industrial composting is usually faster because it runs at higher temperatures, controlled moisture, and aeration. For example, studies and certification guidance commonly describe industrial composting as a thermophilic process operating at around 55–60 °C, with high moisture and oxygen availability. Home composting is normally cooler and less controlled. The home-compostable claims are checked at ambient temperatures (roughly 21–28 °C), with longer allowed time windows than those in industrial composting schemes.

Main Biodegradable and Compostable Polymers Used in AM: In AM, most 'biodegradable printing' is still dominated by PLA and PLA-based blends, but the material set is broader than PLA alone. PLA is the most common biodegradable polymer in FFF/FDM because it prints easily and is widely available. PLA's biodegradation in compost tends to be much faster under thermophilic composting conditions than in soil or water, as shown in broad PLA biodegradation reviews. A key limitation is that printed PLA products may not degrade quickly in cooler, less controlled environments, unless the conditions support hydrolysis and microbial assimilation [13].

Polyhydroxyalkanoates (PHAs) (such as PHB and PHBV) are often highlighted because they can biodegrade in a wider range of environments than PLA, depending on grade and conditions. Soil and marine biodegradation mechanisms for PHAs are widely reported, including microbial depolymerase activity. There is also active work on PHA blends for FDM to improve processability and expand applications beyond biomedical use [14]. Poly (butylene succinate) (PBS) is an aliphatic biodegradable polyester with good processability in conventional plastics processing. Reviews indicate that PBS properties and biodegradation rate can be tailored through molecular design [15]. In AM, PBS is considered a candidate filament polymer partly because of its relatively low melting point (below about 120 °C) and ductility [16]. PLA–PBS blends are also explored to improve performance relative to neat PLA in certain print settings [17]. Polycaprolactone (PCL) is widely used in biomedical research because it is biodegradable and can be processed into 3D-printable forms for scaffolds. Reviews indicate substantial research activity on PCL printing for tissue engineering [18]. However, PCL degradation may occur more slowly than many expect, which is important for compostability claims and for its long-lasting presence in certain environments [19].

Many commercial 'compostable plastics' are blends rather than single polymers. This is important for AM because blending is a practical way to balance printability, toughness, and end-of-life behaviour. For example, PLA/PHB systems can show biodegradation under industrial composting and, for some formulations, even home-composting ranges, demonstrated in controlled tests and in a municipal composting plant setting [20]. Adding biomass fillers (such as algae residues or wood-based fillers) can alter biodegradation rates and mechanical behaviour, sometimes accelerating degradation by supporting microbial growth or altering water transport [21].

A major AM-specific point is that the manufacturing method changes the structure. This is not only about shape; it also affects polymer morphology, porosity, and water-uptake pathways. PLA biodegradation in compost is strongly affected by crystallinity. A controlled composting study on injection-moulded PLA shows that higher crystallinity reduced the degradation rate because the structure is denser and less permeable to water diffusion and enzymatic attack [9].

Research on accelerating PLA degradation also notes that amorphous regions typically permit faster water diffusion, so crystallinity directly relates to hydrolytic degradation behaviour [22]. This matters in AM because printing conditions (cooling rate, bed temperature, annealing, and raster pattern) can alter crystallinity and orientation, which, in turn, affect the biodegradation rate. AM parts can have different surface-area-to-mass ratios than moulded parts. A recent study comparing 3D-printed PLA and injection-moulded PLA under controlled composting found that the 3D-printed specimens disintegrated faster, which the authors linked to their higher surface area and greater water sorption [23]. At the same time, thick-walled parts and high-mass parts can be slow to disintegrate in real compost. The effect of thickness is discussed in polymer degradation research and is a major reason why thin-film results should not be assumed for thick products. In AM, additives are common: colour masterbatches, lubricants, impact modifiers, and fillers (wood, starch, algae, and natural waste fibres). These can change water diffusion, microbial colonisation, hydrolysis rate, and fragmentation behaviour. One controlled composting study on PLA/algae biocomposites reported increased biodegradation when algae biomass was added and discussed microbial growth and nitrogen effects as contributing factors [24].

Biodegradable or compostable polymers can improve sustainability, but only in specific use cases.

The compostable AM polymers make the most sense when the product is likely to be contaminated with food/organic waste, and recycling is not realistic, or the product is part of a controlled organic waste stream where certified compostables are accepted and processed properly. Composting converts polymer carbon mostly into CO_2 (and some biomass). That means losing the material value. For durable printed parts (such as fixtures, housings, enclosures), reuse and mechanical recycling can be better sustainability options than composting, if the polymer can be collected cleanly. The study comparing 3D-printed vs moulded PLA also examined mechanical recycling cycles and linked them to changes in degradation behaviour, showing that 'recycle then compost' is a possible pathway in some scenarios.

Operational sustainability improves when biodegradable and compostable polymers are treated as engineered systems, not as simple 'green replacements'. This system includes the material, the printing process, and the end-of-life route.

In additive manufacturing, the first step is to decide the disposal method. If industrial composting is planned, materials should meet standards such as ASTM D6400 or EN 13432. These standards confirm breakdown under controlled, high-temperature composting. For home composting, materials must be tested at lower temperatures. Simply having industrial certification isn't enough. The thickness and mass of a part are important factors. These factors significantly affect the results of compostability tests. Thin films can degrade quickly, while the breakdown of thick printed parts can take a long time. This can happen even when the material is certified. Printing conditions also affect degradation. Print temperature, cooling rate, and layer structure change crystallinity and internal shape. These changes affect how water enters the material and how microbes act on it. Therefore, the compostability of the final printed object should be evaluated rather than that of the raw resin.

5.3 SUSTAINABLE METAL POWDERS

Metal powders (fine particles of iron, aluminium, nickel, titanium, etc.) are key to powder metallurgy and additive manufacturing (AM). These powders enable near-net-shape parts with minimal waste, but their production is energy- and resource-intensive. Recent literature shows a strong industry and academic push towards sustainable powders that use recycled feedstocks, clean energy, and closed-loop recycling. Major powder makers (e.g. Höganäs, EOS, 6K, Continuum) have set carbon targets and are developing low-carbon processes [25, 26]. For example, Höganäs AB (a leading iron-powder producer) targets net-zero emissions across its entire value chain by 2037, planning to switch to renewable electricity and bio-derived process carbon (biochar) [26].

5.3.1 Production Methods and Feedstocks

Metal powders are manufactured by several methods. Gas atomisation is most common for high-quality powders: molten metal is forced through a small nozzle and hit by a high-speed inert gas jet (argon or nitrogen), breaking the melt into fine droplets that solidify into spherical powder. Water atomisation (high-pressure water jets) is used especially for steels and other ferrous alloys – it yields irregular particles but has a higher yield and lower gas use. Other processes include plasma atomisation (metal vaporised by plasma arc), rotating electrode (PREP) for refractory alloys, centrifugal atomisation, as well as mechanical methods (milling or mechanical alloying) and chemical routes (e.g. metal-carbonyl decomposition). In all cases, the powder is then sieved and classified to obtain a narrow particle size distribution suitable for AM or powder metallurgy [27].

Feedstock Sources: Powders can be made from primary metals or recycled material. Virgin feedstock (alloy ingots, wrought billets) is melted and atomised for high-purity needs. Importantly, recycled scrap is widely used: machining chips, worn parts, or end-of-life components can be re-melted. For example, aerospace industry partnerships recycle millions of pounds of high-grade alloy scrap – Boeing and Alcoa recycle ~8 million lbs of aluminium scrap into new aerospace-grade stock each year, and Rolls-Royce's 'Revert' programme reclaims >95% of nickel-based superalloy scrap for re-melting [28]. Many suppliers now use certified scrap: one survey notes that 'scrap is the new gold' in metal powder production, with large powder makers relying on steel, nickel, aluminium, and copper scrap to lower costs [29]. Some advanced processes (e.g. Continuum's Greyhound M2P) even take 100% recycled alloy as input. Other feedstocks include direct-reduced iron (for steel powders) or chemical precursors (e.g. nickel carbonyl for Ni powder). Proper sorting and quality control (spectrographic sorting, cleaning, etc.) are needed to prevent contaminants when using mixed scrap [29, 30].

Environmental Impacts and Life Cycle Analysis (LCA): Metal powder production is energy-intensive, so energy use drives its environmental footprint. Life cycle studies consistently find that melting/atomisation energy dominates greenhouse gas (GHG) emissions [27]. For example, a recent LCA of gas-atomised powders found on the order of 20 kg CO_2-equivalent per kg of powder for titanium and nickel alloys. In that study, vacuum induction melting (VIGA) and electrode induction (EIGA) atomisation had similar footprints (~20 kg CO_2/kg), with the high melting point of Ti making Ti-alloy powders slightly more carbon-intensive than Ni-based ones. By contrast, analyses of lighter metals report lower footprints: one study of aluminium gas atomisation found roughly 4.6–6.7 kg CO_2/kg under optimised conditions. These differences reflect alloy properties and process specifics, but all agree that electricity and inert gas use are key factors. For example, if argon gas is preheated before atomisation, it actually increases environmental impact (due to extra energy) compared to using cold gas [31].

Notably, recycling scrap feedstock yields dramatic reductions in environmental impact. LCAs show that using recycled material (and renewable energy) instead of virgin metal can cut emissions by 90–99%. In one study, producing Inconel 718 powder from metal scrap (instead of primary alloy) slashed both energy use and carbon footprint by over 90%. Similarly, aluminium is highly efficient to recycle: secondary (scrap-based) Al requires only ~5% of the energy of primary production, resulting in about 0.6 kg CO_2/kg for recycled Al (vs. up to 14.4 kg CO_2/kg for primary aluminium) [32]. Continuum Powders reports that its plasma-based recycle process (using green power) cuts GHGs by ~99.7% compared to conventional virgin Ni powder production. In short, substituting recycled feed and clean energy can reduce powder production's carbon footprint by one to two orders of magnitude.

Key LCA insights: Energy use is the main contributor (mainly electrical power for melting). Depending on metal and process, footprints span ~5–20 $kgCO_2/kg$. Gas atomisation yields are often low (e.g. only ~30% of the melt ends up in the target size range), so multiple re-melts or re-classifications may be needed. Future atomisation technologies promise large efficiency gains – a prospective LCA suggests next-generation EIGA and plasma atomisers could halve electricity use and reduce argon consumption by up to 98%.

Recycling, Reuse, and Circular Economy Practices: To embed metal powders in a circular economy, industry and researchers emphasise recycling and reuse at every stage. Industrial recycling loops: As noted, aerospace companies recycle superalloy scrap in closed loops. Automotive and tooling industries similarly reclaim steel and aluminium scrap. Powder reuse in AM: In powder bed AM, unsintered 'revert' powder is routinely sieved and reused for a few cycles; this reduces waste (typical reuse rates ~85–95%) but eventually degrades powder quality (oxidation, satellites). Simple blending or sieving can extend powder life, but impurities (O_2, N_2 uptake) accumulate.

Advanced Recycling Methods: The proposed new processes go beyond sieving. Cold-spray techniques have been demonstrated to recycle metal waste, for example, by cold-spraying screened overspray and machining waste (Ti, Al, Cu, 316L stainless) into coatings, which can recover ~70% of the powder and reduce raw material costs by ~30%, illustrating circular manufacturing. Specialised systems (e.g. Amazemet's Powder2Powder) clean, re-melt and re-atomise spent powder, restoring morphology and chemistry. Companies like Continuum also apply strict sourcing and melting steps: they start with certified aerospace alloy scrap, and their Greyhound M2P plasma atomiser (with cold hearth refining) converts it into virgin-grade powder in one step.

Quality Control: Because recycled powders can contain contaminants (oxides, moisture, tramp elements), effective QA is critical. Industry standards (especially in aerospace/medical applications) require tracking oxygen content, particle size, and chemistry during recycling. Powder tracking and testing (e.g. batch certificates, oxygen analysis) are best practices to ensure safety. In sum, these recycling strategies – reclaiming scrap, reusing AM revert, and fully re-melting used powder – keep metal flows closed-loop and greatly reduce the need for virgin ore [30].

Low-Carbon Manufacturing Practices: Beyond recycling, low-carbon manufacturing focuses on reducing emissions during production.

Renewable energy and fuels: Powder producers are shifting to green power and fuels. Höganäs plans to procure 100% fossil-free electricity and to replace coal/coke with biochar or biogas by 2030. Other firms similarly aim to electrify melting and use renewables; for example, R&D is underway to use hydrogen to melt or reduce metal feed (especially for steel powders) and to integrate solar or wind power into plant operations.

Energy-Efficient Processes: New atomisation technologies are also beneficial. 6K Additive's microwave plasma atomiser and Amazemet's ultrasonic atomisation can melt scrap quickly with better thermal efficiency than traditional crucibles, reducing energy consumption. These methods (and others like plasma rotating electrode) often require less inert gas or allow gas recirculation, cutting indirect emissions.

Cleaner Raw Materials: In some cases, sourcing low-carbon raw inputs can help. For example, 'green steel' from hydrogen-reduced iron or recycled Direct Reduced Iron could feed steel powder production with far less CO_2 than blast-furnace iron. Material and alloy design also play a role: choosing alloys that tolerate higher scrap content or that require less melting energy supports sustainability.

Industry Examples Abound: EOS has introduced 'Responsible' aluminium AlSi10Mg made from 100% recycled Al. Continuum's LCA (above) shows how combining recycled feed with green power yields near-zero emissions. Even so, the sector recognises that energy decarbonisation is critical. An LCA survey notes that energy consumption dominates powder production impacts, so any measures (renewables, efficiency) will have outsized benefits. Many companies set science-based targets: Höganäs's SBTi-approved roadmap, for instance, includes strict scope-3 reductions and end-to-end net-zero.

5.3.2 Best Practices, Technology Gaps, and Policy Recommendations

Best Practices: Proven measures include maximising recycled content (while ensuring purity), using energy-efficient melting (e.g. induction furnaces on green grids), and closely tracking powder quality. Leading powder producers now establish product carbon footprints (PCFs) for each grade – for example, Höganäs publishes cradle-to-gate CO_2/kg for its powders under ISO 14067. This transparency allows engineers to make material choices based on emissions. Another best practice is closed-loop design: collaborating with scrap suppliers and customers to reclaim and reuse spent material (as done in aerospace recycling programmes and turbine blade reclaiming). In AM operations, strict powder-handling protocols (inert storage, minimal exposure) prolong powder life and reduce waste.

Technology Gaps: Challenges remain. LCA data on powder production are still sparse: most LCA databases lack dedicated processes for atomisation, making accurate footprinting difficult. We need more lifecycle studies covering diverse alloys and emerging processes. On the technical side, recycling powder without quality loss is nontrivial – contaminant removal (oxides, tramp elements) often requires energy-intensive re-melting or refining. Powder reuse in AM is limited by powder degradation (oxygen uptake, satellites); better real-time monitoring and conditioning between reuse cycles are needed. High requalification costs are also a barrier. Moreover, scaling up new atomisation methods such as microwave, plasma, and ultrasonic technologies for large-scale industrial use remains a challenge. These systems must also be shown to be reliable according to the strict standards of the aerospace and medical fields.

Policy Recommendations: To accelerate sustainable powders, governments and industry bodies should encourage recycled content and low-carbon manufacturing through incentives and standards. For example, public grants or tax breaks for recycled powder use (as already offered for recycled aluminium) can shift the economics [33]. Mandates or procurement criteria could require minimum recycled content or maximum carbon footprints for critical applications. Expanding carbon pricing to include metal feedstocks would make recycling more competitive. Standardisation bodies should support the labelling of powder footprints (as Höganäs does) and develop guidelines for recycled powder qualification (tracking chemistry/oxygen). Public funding for R&D on green metal powder technologies (e.g. hydrogen-based processes, efficient atomisers) is also needed. Höganäs underscores the need for policy backing. Transforming the metals industry will demand incentives from various governmental tiers, coupled with favourable regulations.

The creation of sustainable metal powders is fundamentally dependent on the incorporation of established methodologies. These include the utilisation of scrap feedstock, the application of efficient atomisation techniques, and the implementation of powder recycling protocols, all while concurrently enacting robust decarbonisation strategies and supportive policy frameworks. Consequently, by fostering a circular economy within metal applications and promoting cleaner production methodologies, the industry can substantially reduce its environmental impact.

5.4 CONTINUOUS CARBON FIBRE COMPOSITES

Continuous fibre composites combine very strong fibres (often carbon) with plastics to make lightweight, stiff parts. In this approach, continuous strands of fibre are placed in the print head along with the polymer. This creates parts whose stiffness and strength mainly come from the fibres. As a result, even a small amount of fibre can greatly increase the part's load-bearing ability compared to the plastic alone. For example, adding continuous carbon fibres to a nylon matrix can create a material with a much higher specific stiffness (stiffness relative to weight) than the same plastic without fibres [34, 35, 36].

Natural fibres like flax or hemp can also be used for lower-strength parts. These bio-based fibres have good stiffness and are lighter than glass or carbon fibres. In practice, printers (such as those by Markforged or Anisoprint) lay carbon fibre filaments along stress paths or fill zones with a 'cross-hatch' fibre pattern. The main benefit is weight reduction. A bracket made with continuous carbon fibre plastic might weigh 50–70% less than a similar metal part, while carrying comparable loads. This advantage is particularly significant in fields such as aerospace and automotive engineering, where even a single kilogram of weight reduction can improve fuel economy.

Sustainability considerations: Despite their good performance, continuous carbon fibre composites present challenges for sustainability. The fibres themselves are not biodegradable and manufacturing them takes a lot of energy (making carbon fibre yarn or tape is very energy-intensive). At the end of life, recycling is difficult. Thermoplastic composites (with re-meltable plastic) can be mechanically recycled by shredding and re-melting, but this usually yields weaker short-fibre material (no longer continuous), which is used for low-grade parts or fillers.

Thermoset composites (with uncured resin): They can't be re-melted. They are typically recycled by thermal processes (pyrolysis) that burn off the resin and recover the fibre. This recovers some of the carbon fibre (often with ~70–80% of its original strength) but is costly and not widely practised. In all cases, components must be collected and processed in special recycling facilities. There is currently no standard 'industrial composting' for carbon fibre parts. As a result, end-of-life strategies usually focus on the reuse of parts or careful recyclability: design parts for disassembly, so fibres or components can be separated from other materials, avoid mixing incompatible polymers (multi-material prints), since multi-material composites are very hard to separate and recycle, and natural fibre composites (flax/hemp) have an advantage in end of life. Because they use plant fibres and a biopolymer matrix (such as PLA), they can often be processed like other bioplastics. In some cases, they may biodegrade under composting conditions or at least produce recyclable by-products. This does not make them fully compostable, but it makes recycling easier than for pure carbon fibre. For instance, a PLA-flax printed part can, after use, be ground into mulch or composted in industrial facilities, something you cannot do with pure carbon fibre parts.

Best practices: Fibres should be used selectively, only in regions that require enhanced structural strength. The blanket application of fibre throughout a component results in resource wastage and complicates recycling processes. Proper fibre alignment is essential, with fibres oriented along the primary stress trajectories to optimise performance. This approach ensures the maximisation of the material's strength-to-weight ratio [34, 37].

5.5 3D PRINTING WITH RECYCLED MATERIALS

3D printing (additive manufacturing (AM)) is increasingly viewed as a route to a circular plastics economy. Global plastic waste is enormous – only ~9% is currently recycled – while the 3D printing market is booming (projected from ~$15.1B in 2021 to $44.5B by 2026) [38]. By feeding waste plastics into printers, the industry can cut raw material use and emissions [39]. Recent reviews confirm that many recycled filaments match the performance of virgin plastics; for example, 3D-printed parts made from recycled plastic have shown comparable performance to virgin materials [40]. Key polymers like PLA, PET, and ABS are now repurposed, often with additives or fillers to boost strength. Overall, experts emphasise careful processing and adherence to standards (ASTM/ISO guidelines) as recycling scales up in AM [40].

5.5.1 Common Recycled Feedstocks and Their Properties

Many standard 3D printing plastics have recycled variants. For example, Filamentive (a UK filament maker) uses post-consumer PET bottles and industrial PLA scrap to produce reusable PET and PLA filaments. Typical recycled filaments include:

Polylactic Acid (PLA): PLA is a plant-based bioplastic (corn, sugarcane), with a melting point of ~150–180 °C.. It prints easily with low warping. Recycled PLA generally retains good strength: one study found only ~10–15% drop in tensile strength after one recycle cycle. Some drop in molecular weight does occur after repeated processing.

Polyethene Terephthalate (PET/PETG): It can be derived from beverage bottles. When recycled (often to PETG for 3D printing), it has high durability and chemical resistance. Recycled PET maintains good mechanical performance. For example, recycled PET filaments are commonly used in packaging and in upcycling textile waste [39].

Acrylonitrile Butadiene-Styrene (ABS): It is a tough engineering plastic. Post-consumer ABS (e.g. from housings or bottles) can be made into filament, but it tends to warp and emit fumes, so printing requires an enclosure and ventilation. Studies show recycled ABS works in FFF printers, though often with slightly altered settings.

Polyolefins (PP, HDPE, and LDPE): Commodity plastics (e.g. milk jugs, containers). These have high shrinkage, so recycled PP/PE filaments require strict control (heated beds/chambers). When made correctly, these materials create parts that are both lightweight and resistant to chemicals.

Other Polymers: Some higher-performance plastics are also recycled. For instance, polycarbonate (PC) from used bottles is recycled into 3D filament. (Polymaker's PC-r filament from PC bottles showed strength slightly above virgin PC.) Thermoplastic polyurethane (TPU), nylon (PA), and speciality thermoplastics (PPS, HIPS) can likewise be reprocessed, often blended with virgin resin or fibres.

Filled/Composite Filaments: Many recycled filaments include natural or mineral fillers. Agricultural or wood flour waste, carbon/graphite, and even recycled carbon/glass fibres (e.g. from wind turbine scrap) are added to stiffen recycled polymers. These composites raise stiffness and dimensional stability, further supporting the upcycling of biomass or industrial wastes.

5.5.2 Recycling Process and Printer Adaptations

Typical recycling-to-print workflows involve sorting, cleaning, shredding, extruding, and printing. A common five-step process is (1) sort plastics by type, (2) wash and dry them, (3) shred/granulate into flakes, (4) melt and extrude into filament or pellets, and (5) print with those filaments. Commercial equipment, such as filament extruders like the Filabot and 3devo's Filament Maker, handles the melting and winding processes. Several critical factors come into play:

Feedstock Preparation is essential. The polymer types such as PLA, ABS, PET must be accurately identified. Mixing incompatible plastics (PET vs. PLA) is not advisable. Clean and dry scrap thoroughly (3devo's shredder + desiccant dryer setups are common) to avoid contaminants. The quality of filaments begins with the use of consistent pellets that are free from any contaminants.

Extrusion Controls: The melt temperature must comply with the polymer's specifications. For instance, ~180–220 °C for PLA and 220–250 °C for ABS. During extrusion, precise diameter control is critical – even a 0.05 mm variation can cause under- or over-extrusion. Modern filament makers use laser gauges and closed-loop control to maintain uniform 1.75 mm (or 2.85 mm) filament. Some setups allow co-extrusion or blowing air to cool the strand evenly [41].

Printer Settings: Recycled filaments often require some fine-tuning of print profiles to achieve the best results. Users typically dry the filament beforehand (for example, PLA/PETG at ~45–60 °C) to eliminate moisture. Print speeds are kept moderate, and flow/extrusion multipliers may be lowered to compensate for slight density differences. A heated build plate (50–100 °C) helps improve adhesion for many plastics. For warping-prone materials (ABS, PP), using enclosures and maintaining a consistent bed temperature is strongly recommended. Adding adhesion aids (PEI sheets, glue stick, hairspray) is a best practice to improve first-layer bonding [42].

5.5.3 Case Studies and Commercial Solutions

Filamentive (UK): A leading filament brand producing 100% recycled filaments. Filamentive's 'ONE' PET filament is made from post-consumer recycled PET bottles, and their PLA filament uses pre-consumer PLA waste. They emphasise ISO-14021 'recycled content' standards and conduct rigorous quality testing.

3DPW Limited (UK): An SME offering 3D printer waste collection and recycling. In a case study, 3DPW shredded and dried PLA (and trials of ABS, PETg) waste from client printers and extruded it into new filament. They demonstrated prototypes printed on this recycled filament, aiming to close the loop on lab/school 3D printing waste.

Covestro/Polymaker (Global): A chemical-materials partnership. Covestro supplied recycled polycarbonate from ~19-litre water bottles, which Polymaker processed into PC-r filament. This industrial-scale example delivers a high-performance filament that meets sustainability criteria (lower CO_2 footprint and industry certifications). Notably, the pure source stream (single-origin bottles) eliminated sorting issues.

Prusa Research (Czech Rep.): A desktop 3D printer company with in-house recycling. Prusa produces 'Prusament Recycled' PLA and PETG from their own production scrap (not failed prints). Their LCA-based analysis shows dramatic savings: 1 kg of recycled PETG emits 2.55 kg CO_2 vs 5.85 kg for virgin PETG (a ~56% reduction). PLA-R recycled spools showed a similar ~57% cut in embodied carbon. These spools are sold at the same price as virgin filament, demonstrating commercial viability.

3devo (Netherlands): Manufacturer of desktop recycling machines. 3devo sells filament extruders, plastic shredders, and dryers tailored for small-scale recycling. Their machines allow schools, labs, and businesses to create high-quality recycled filament on-site. For example, their 'Filament Maker' extruder can turn shredded PLA bottles into filament with tight temperature and diameter control.

Industry and Maker Initiatives: Companies like HP and BASF/Covestro are testing recycled/bio filaments. HP's recent LCA found that using 70% recycled PET in a filament reduced its carbon footprint by ~31.6%. Community projects (Precious Plastic, RecycleBot, and RePLAy 3D) demonstrate local recycling: DIY filament extruders and mail-in recycling services aim to close the loop at the user level. Many global efforts (e.g. desktop pellet extruders, recycling programmes at Fab Labs) are emerging to turn local plastic waste into feedstock.

Transforming plastic waste from 3D printing into usable feedstock is technically possible and offers environmental benefits. On the other hand, this process requires careful planning and precise execution. The community suggests thorough sorting, strict quality control, and a gradual approach to using it. For example, they recommend starting with a mixture of 20–50% recycled material and new resin. As more companies and labs share data (e.g. LCA results, processing parameters), best practices are emerging. Using recycled filament in 3D printing significantly reduces both carbon emissions and material waste. This makes it a worthwhile approach for more sustainable manufacturing practices.

5.6 HYBRID AND MULTI-MATERIAL PRINTING

Hybrid and multi-material additive manufacturing combine diverse technologies and materials to expand what can be printed within a single workflow. Hybrid manufacturing (HM) integrates additive processes with subtractive (machining) or other methods on a single machine or cell [43]. By contrast, multi-material AM allows different materials (polymers, metals, elastomers, etc.) or colours to be printed simultaneously or sequentially to create functionally graded or composite parts. Both approaches aim to leverage the strengths of each process or material: HM marries AM's geometric flexibility with machining's precision and surface finish, while multi-material AM offers new design freedom (e.g. multi-colour aesthetic parts or integrated rigid-soft features) without assembly [44].

5.6.1 Hybrid Additive/Subtractive Manufacturing

Hybrid machines combine an additive head (e.g. laser cladding or wire-DED) with a subtractive head (e.g. Computer Numerical Control (CNC) milling) on the same platform. This single-setup approach avoids moving parts between machines, reducing wasted time and error. For example, combining powder bed fusion (PBF) with in situ milling yields complex metal parts with fine tolerances: the laser builds the geometry layer by layer, and a milling tool smooths faces or adds features [43]. Recent reviews report that such hybrid AM systems can halve build time and improve surface roughness by ~70%, while keeping dimensional errors to within fractions of a millimetre [44]. For instance, Wire Arc AM (WAAM) paired with CNC milling produces larger parts faster and with better precision, and combined laser-based powder deposition plus machining has been shown to enhance feature accuracy and material utilisation.
Hybrid setups fall into broad categories:

Separate Machines: An AM machine followed by transfer to a CNC machine. It is most common today but involves extra handling.
Cellular Hybrid: Dedicated stations where a robot or sliding table allows AM and subtractive manufacturing (SM) heads to work on one part in sequence.
Integrated Hybrid: A single multi-axis machine that swaps tools (or has multiple heads) to add and remove material without re-fixturing.

Fully integrated hybrids can automatically switch between deposition and milling heads in a single cycle, greatly increasing throughput and reducing manual handling. Many new hybrids are based on directed energy deposition (DED), since DED can both build new parts and repair or add to existing components. For example, a DED head on a 6-axis robot can weld metal onto a part while another tool machines it, enabling on-demand repair of turbine blades or die inserts [43].

However, hybrid AM also brings new challenges. The combined systems are complex and costly to install and maintain. Thermal management and process synchronisation between the additive and subtractive tools must be carefully controlled to avoid distortions or mismatches. Material interfaces (e.g. between additively deposited layers and machined surfaces) require attention to avoid stress or defects. In practice, multistep calibration and robust process monitoring are needed to realise the full benefits [44]. Despite these hurdles, hybrid AM holds strong promise for aerospace, biomedical, and tooling applications by enabling functionally graded materials (FGMs) and interior channels that would be difficult or impossible with one process alone.

5.6.2 Multi-Material Additive Manufacturing

Multi-material AM refers to printing with two or more distinct materials in one build. This can include different polymers (e.g. rigid vs. elastomeric), polymers and ceramics, metal alloys, or even polymer-metal combinations [45]. By assigning materials spatially, designers can create multi-colour parts, or parts with zones of different stiffness, biocompatibility, or electrical conductivity. For example, Polymer Jetting (PolyJet) systems (e.g. Stratasys J-series) have multiple print heads loaded with rigid and flexible photopolymers. They blend droplets of 'Vero' and 'Tango' resins to achieve custom rubber-to-plastic mixtures or tens of thousands of colours [46]. Similarly, Binder Jet printers can print parts with two powders by selectively jetting different binders, enabling coloured gypsum models or sandwich structures.

Key enabling technologies include multi-nozzle extrusion FDM (using sequential filament changes or multiple extruders), material jetting (PolyJet or MJ handles voxel-level colour), and hybrid techniques like embedding pre-made parts into prints. Recent innovations have also explored AM of functionally graded materials (FGMs), where a gradient of material properties is built in; for instance, digital design of polymer-derived ceramic composites printed layer by layer. These advancements allow a single print to serve multiple functions or aesthetic purposes in one shot.

Despite these possibilities, material compatibility is a significant limitation. Not all materials bond well; FDM printers typically use similar thermoplastics (PLA, ABS, TPU, and PETG), so adhesion is good, but exotic mixes (e.g. PLA and TPU) can warp. PolyJet polymers are engineered to blend but still have a limited palette of compatible resins (rigid, digital ABS, rubber, and clear). Metal multi-material (different alloys) requires careful melt pool control and is often limited to successions, e.g. steel–copper in directed deposition. Mismatched thermal or cure profiles can cause delamination at interfaces, so design for multi-materials requires simulation and testing.

5.6.3 Performance and Material Compatibility

The performance of hybrid or multi-material systems depends on hardware and materials. For example, high-end PolyJet 3D printers achieve 14 μm layer thickness (0.00055') for extremely smooth surfaces. They can blend hundreds of thousands of colour/material combinations in a single part. By comparison, FDM systems typically layer ~100–200 μm, and their multi-material resolution is bounded by nozzle diameter. Multi-material PolyJet parts can have Shore hardness variation and RGB colour per voxel, enabling lifelike prototypes. However, these photopolymers have modest mechanical strength (such as nylon/ABS level) and limited high-temperature resistance. Metal–metal hybrids (such as DED/SLM machines) can produce full-density alloys, but multi-phase metal prints often have weaker material interfaces or porosity that must be heat-treated.

Performance metrics often cited include build rate, accuracy, and surface finish. Studies show that adding a milling step can improve surface roughness by ~70% and maintain +/-0.5 μm accuracy. In one survey, hybrid AM reduced overall cycle time by up to 50% compared to sequential AM+machining. On the multi-material side, key metrics are material compatibility (tested by bond strength), colour accuracy, and functional gradation (smoothness of property transition). Quantitative benchmarks are still emerging, but research emphasises characterising interfaces (e.g. tensile strength across a polymer–polymer or metal–polymer boundary) [44].

5.6.4 Applications and Case Studies

Hybrid and multi-material AM are already showing impact in many fields. In aerospace, hybrid DED+milling machines are used to repair turbine blades, add material where needed, and then machine to spec. Hybrid processes can yield parts with graded cooling channels or tailored stiffness – for example, printing a part with metal matrix composites in high-stress zones and pure metal elsewhere. Multi-material printing can produce complex injection-moulding masters in one step (rigid core with soft overmold), and even control local properties in composite moulds.

In biomedical and medical devices, hybrid machines enable printing and machining of medical implants with precise fit and surface polish. Multi-material AM enables prosthetics or orthoses that combine hard plastic frames with integrated soft pads or silicone-like cushions, printed seamlessly. One study notes the promise of functionally graded AM for orthopaedic implants, blending stiff and compliant materials to better match bone and cartilage tissues (though clinical examples are still emerging) [45]. Full-colour multi-material PolyJet printing is used in anatomy models and surgical planning to replicate the appearance of tissues.

In consumer and design prototyping, the use is widespread. Designers use multicolour/multi-material prints to create concept models with true finished-look aesthetics (e.g. a single print containing both a model car's translucent headlights and matte bumper). Hybrid manufacturing is less common in consumer hardware today, but it is enabling new metal products with internal lattice structures plus high-quality surfaces in one build. Automotive and aerospace companies have demonstrated multimaterial 3D printed ducts that integrate hard structural sections with rubber-like seals, and hybrid AM parts with optimised stiffness.

5.6.5 Standards, Guidelines, and Research Gaps

Fundamental definitions for AM are covered by ISO/ASTM 52900:2021 (e.g. listing 'Material Extrusion, Material Jetting' etc.) [43], but no single international standard yet governs multi-material or hybrid printing. Existing AM standards focus on processes or material classes. The field would benefit from guidelines on the qualification of multi-material parts (e.g. how to validate a graded material). Equipment standards (like ISO/ASTM 52904 for laser powder bed) are emerging for hybrid setups, but hybrid machines often rely on proprietary controls and are evaluated per use-case. Current research gaps include: multi-material interface modelling and testing methods, multi-physics simulation for hybrid thermal fields, and design tools for creating functionally graded or composite parts. For example, one review calls out the need for 'design tools' to handle building-scale multi-material structures (e.g. concrete/plastic blends). Real-time monitoring and adaptive control for hybrid machines is also underdeveloped; integrating sensors to detect build defects on the fly remains a challenge.

Looking forward, research should focus on: new material pairings (e.g. polymer–metal composites), improved adhesion techniques, standardised test coupons for hybrid builds, and expanded multi-axis hybrid platforms (e.g. robots that can climb on builds). Both industrial and academic sources stress the importance of cross-disciplinary optimisation: combining machine, material, and design innovation. The vision is a future in which intricate structures are both 3D-printed and finished on a single machine. This would involve using materials that are carefully combined, with different compositions to improve strength, functionality, and appearance.

5.6.6 Potential Recommendations

It is recommended to develop open-source benchmarks for hybrid additive manufacturing, such as standard 'trimmed' geometries. To expand material data sheets to include multi-material adhesion strengths and collaborate to establish multi-material standards; for instance, ISO committees are developing standards for functionally graded materials. To accurately evaluate these technologies for production, more case studies focusing on end-use parts, rather than just prototypes, are necessary. By addressing these gaps, hybrid and multi-material printing can become vital tools for mass customisation and high-performance manufacturing across various industries.

5.7 IN SITU MONITORING AND PROCESS CONTROL

Additive manufacturing (AM) processes now increasingly embed in situ sensors to monitor each layer as it is built [47]. These sensors (thermal cameras, photodiodes, acoustic and vibration sensors, etc.) capture real-time data on melt pool temperature, layer geometry, machine vibrations, etc. By analysing this data in real time, systems can detect defects (e.g. porosity, delamination, overheating) and adjust process parameters (laser power, feed rate, etc.) to correct them. In practice, in situ monitoring improves quality control and reduces scrap; for example, NASA developed an inspection system that combines infrared and visual cameras to compare the actual build with the design in real time. In material extrusion AM (FDM/MEAM), researchers have shown that embedded sensors (accelerometers, encoders, pressure transducers, cameras, etc.) can continuously track printer performance and material flow, supporting predictive maintenance and fewer failed prints [48].

5.7.1 Key Sensors and Data Types for Real-Time Monitoring

In additive manufacturing (AM) applications, a variety of sensors are employed to collect real-time data pertaining to the build process [49, 50]. Typical examples encompass:

Thermal/Infrared Sensors: High-speed pyrometers and infrared cameras are utilised to ascertain the temperature of the melt pool or the build bed. These provide temperature maps that correlate to melt pool size and stability. For instance, NIST demonstrated a 22,500 fps colour camera for two-colour melt pool thermography.

Optical (Visual) Cameras: Visible-spectrum cameras (often high-speed) capture layer geometry, weld/bead shape, and powder bed conditions. Visual imaging is used for melt pool monitoring or layer-wise dimension checks. Some systems even reconstruct an as-built 3D model from layer images.

Photodiodes and Pyrometers are used to measure the power of lasers and electron beams, as well as the intensity of plumes; photodiodes are fast photodetectors. By monitoring laser intensity and reflected/emitted light, the system can infer melt pool health and detect overheating. For example, embedded photodiodes are standard for real-time laser power feedback.

Acoustic Emission (AE) and Vibration Sensors: Piezoelectric AE sensors detect high-frequency sound waves from cracking, delamination, or spatter events. Vibration accelerometers on the machine frame or nozzle can sense mechanical faults (motor stalls, clogs) and some part quality issues. Research has shown AE can pick up layer detachment and material jams in FDM.

Load/Pressure/Force Sensors: In extrusion-based AM, pressure transducers and load cells in the nozzle or filament drive monitor melt pressure and motor torque. These rheological measurements reveal changes in polymer viscosity or flow that may indicate feedstock issues.

Spectroscopy: Optical emission spectroscopy (OES) or laser-induced breakdown spectroscopy (LIBS) of the plume can identify alloy composition and contamination during metal AM. Such spectral data ensures material homogeneity.

Strain Gauges and Thermocouples: For large systems (e.g. powder bed printers), strain gauges on the build plate can infer residual stresses, while embedded thermocouples track substrate temperature. For example, one NIST/UNC-Charlotte project instrumented the build plate with strain sensors and an ultrasonic transducer to monitor acoustic waves through the part.

In some cases, ultrasound (through the build plate), laser profilometers, or X-ray imagers (for in situ CT) are used. Generally, combining multiple sensor types (thermal, optical, and acoustic) is considered best practice, as each mode highlights different defect signatures [51].

5.7.2 Control Algorithms and Feedback Strategies

Collected sensor data must be translated into control actions. Broadly, two strategies are used: feedforward (open-loop) and feedback (closed-loop). In feedforward methods, parameters are set beforehand, often based on simulations or test builds, and are not modified during the build. Conversely, feedback control constantly compares sensor measurements to desired targets and adjusts actuators in real time.

Proportional-integral-derivative (PID) controllers are a common feedback control method.

The sensor, like a temperature sensor, measures the current melt pool temperature. The PID algorithm computes an error term and adjusts a process variable (e.g. laser power or travel speed) proportionally to that error, its integral, and derivative. For example, Wang et al. (2023) used a high-speed thermal sensor with a 2 kHz PID controller to regulate laser power during LPBF, significantly reducing defects caused by overmelting [52]. Although PID loops are relatively easy to implement, they require careful tuning and might not work as well in the complex, nonlinear dynamics of additive manufacturing.

Model predictive control and adaptive control are more advanced algorithms that are not yet widely used and involve using a process model to predict future states. The model predictive controller (MPC) would then calculate the optimal parameter adjustments over several steps to keep the melt pool stable. For example, the average temperature of the melt pool can be monitored. If the temperature exceeds the designated thresholds, the build process can be temporarily stopped or adjusted.

Statistical process control (SPC) is used to track the sensor output through statistical thresholds. ASTM standards (e.g. E3353) are emerging to guide such in-process SPC for LPBF.

Machine learning (ML)-based control is a rapidly growing trend that uses AI to interpret sensor data and decide on corrections. Simple ML models (neural nets, SVMs) can classify sensor patterns to flag defects. Reinforcement learning (RL) is especially promising: RL algorithms learn a control policy that maximises build quality over time. Gunasegaram et al. note that deterministic rule-based controllers struggle in AM's stochastic environment, whereas ML (especially RL) provides the flexibility to handle diverse defects [53]. In practice, researchers have experimented with RL and inverse models to select adjustments (e.g. altering scan speed or power) in response to observed anomalies.

Data-Driven Optimisation: Some systems use Bayesian optimisation or evolutionary algorithms between layers: they analyse accumulated sensor data and re-optimise the remaining parameter schedule. For instance, after detecting consistent overmelting in small features, a controller might reduce the laser power for subsequent scan lines (feedforward update).

Hybrid Strategies: In many cases, a combination is used, e.g. feedforward for coarse settings (based on feature geometry) plus closed-loop PID fine-tuning. Multiple control loops might run in parallel: one loop holding melt pool width constant, another maintaining build plate temperature, etc. The key is the tight integration of sensing and actuation.

In all cases, controllers operate on high-bandwidth data: modern research testbeds achieve kilohertz update rates (e.g. the LPBF system above modulated power at 2 kHz). Figure sensors are processed by digital signal processors or GPUs in real time. As literature notes, 'feedback control is generally considered more effective and robust' for AM's dynamic thermal loads.

5.7.3 Case Studies of In Situ Monitoring and Control

Wire Arc EBAM (Sciaky IRISS): Sciaky's Electron Beam Additive Manufacturing system uses a patented IRISS® closed-loop control. A suite of sensors (melt pool cameras, plume sensors, etc.) measures deposition geometry layer by layer. Proprietary software then adjusts the electron beam power, wire-feed speed, and CNC motion dozens of times per second to keep the build uniform. This industrial example shows how in situ vision can directly drive closed-loop control of large-scale parts [54].

Laser Powder Bed Fusion (LPBF) – Experimental Closed-Loop: Wang et al. (2023) built a custom LPBF platform with a high-speed IR sensor and feedback controller. They correlated melt pool thermal emission to part dimensional error via small-feature test prints. During actual builds, the controller modulated laser power (up to 2000 updates per second) to keep the emission at a setpoint. This kept heat accumulation in check: parts printed with closed-loop control had significantly lower error and better surface finish than open-loop prints. This study shows that even a simple PID control loop, which adjusts laser power based on thermal sensor data, can significantly improve the quality of the finished product by adapting to changes in feature size [52].

Aerospace (GE Aviation – LEAP Fuel Nozzles): General Electric's aviation sector integrated in situ melt pool monitoring into the selective laser melting (SLM) fabrication of LEAP engine fuel nozzles. By observing the melt pool and layer formation during each laser scan, they achieved uniform fusion. GE's documentation indicates that this methodology facilitated 'consistent production quality and defect reduction'. Furthermore, they achieved a 25% weight reduction, along with substantial cost savings compared with traditional casting techniques. Using on-site data can streamline the certification process and lighten the load on critical safety components [55].

Automotive (BMW – Engine Brackets): BMW has used in situ monitoring for metal brackets and engine components. One case combined thermal imaging and acoustic emission sensors to watch the build. This dual monitoring caught issues like a lack of fusion early, allowing on-the-fly parameter tweaks. BMW observed that part performance improved and build times fell once these sensors were active. This approach underscores the fact that in situ monitoring isn't solely an academic exercise; leading original equipment manufacturers are already implementing it to improve quality and reduce lead times [55].

They used cameras to check if the layers were aligned correctly and infrared sensors to monitor the material's temperature. The real-time feedback enabled adjustments to the scanning parameters, thereby producing implants characterised by a more uniform microstructure and improved dimensional precision. Consequently, in situ monitoring enabled Stryker to confirm that each implant met stringent biocompatibility and tolerance benchmarks, eliminating the need for protracted post-manufacturing inspections [54].

In summary, in situ monitoring and control are rapidly advancing fields. By combining rich sensor data, smart analytics, and fast control, future AM systems will be able to 'self-correct' defects as they occur, yielding higher quality and lower cost. Collaboration between researchers and machine builders – guided by industry standards – will be crucial to turn these best practices into an everyday manufacturing reality.

REFERENCES

1. S. Bhagia, K. Bornani, R. Agrawal, A. Satlewal, J. Ďurkovič, R. Lagaňa, M. Bhagia, C. G. Yoo, X. Zhao, V. Kunc, Y. Pu, S. Ozcan, & A. J. Ragauskas, Critical review of FDM 3D printing of PLA biocomposites filled with biomass resources: Characterization, biodegradability, upcycling and opportunities for biorefineries. *Applied Materials Today*, 24 (2021) 101078. https://doi.org/10.1016/j.apmt.2021.101078
2. K. Mikula, D. Skrzypczak, G. Izydorczyk, J. Warchoł, K. Moustakas, & K. Chojnacka, 3D printing filament as a second life of waste plastics – A review. *Environmental Science and Pollution Research*, 28 (2021) 12321–12333. https://doi.org/10.1007/s11356-020-10657-8
3. A. Romani, L. Perusin, M. Ciurnelli, & M. Levi, Characterization of PLA feedstock after multiple recycling processes for large-format material extrusion additive manufacturing. *Materials Today Sustainability*, 25 (2024) 100636. https://doi.org/10.1016/j.mtsust.2023.100636
4. Y. Jiang, A. X. Serrano, W. Choi, R. C. Advíncula, & H. F. Wu, Advanced and functional composite materials via additive manufacturing: Trends and perspectives. *MRS Communications*, 14 (2024) 449. https://doi.org/10.1557/s43579-024-00620
5. A. Yousaf, A. A. Rashid, R. Polat, & M. Koç, Potential and challenges of recycled polymer plastics and natural waste materials for additive manufacturing. *Sustainable Materials and Technologies*, 41 (2024) e01103. https://doi.org/10.1016/j.susmat.2024.e01103
6. H. Hegab, N. Khanna, N. Monib, & A. Salem, Design for sustainable additive manufacturing: A review. *Sustainable Materials and Technologies*, 35 (2023) e00576. https://doi.org/10.1016/j.susmat.2023.e00576
7. A. Sola, & A. Trinchi, Recycling as a key enabler for sustainable additive manufacturing of polymer composites: A critical perspective on fused filament fabrication. *Polymers*, 15 (2023) 4219. https://doi.org/10.3390/polym15214219
8. T. Peng, K. Kellens, R. Tang, C. Chen, & G. Chen, Sustainability of additive manufacturing: An overview on its energy demand and environmental impact. *Additive Manufacturing*, 21 (2018) 694–704. https://doi.org/10.1016/j.addma.2018.04.022
9. R. Pantani, & A. Sorrentino, Influence of crystallinity on the biodegradation rate of injection-moulded poly(lactic acid) samples in controlled composting conditions. *Polymer Degradation and Stability*, 98(5) (2013) 1089–1096. https://doi.org/10.1016/j.polymdegradstab.2013.01.005
10. A. Faveto, F. Lombardi, P. Chiabert, & F. Segonds, A circular approach to foster additive manufacturing early design stages sustainability: A methodological proposal. *International Journal on Interactive Design and Manufacturing (IJIDeM),* 18 (2023) 815–826. https://doi.org/10.1007/s12008-023-01577-1
11. International Organization for Standardization. *Specifications for Compostable Plastics (ISO 17088:2012)* (International Organization for Standardization, 2012).

12. International Organization for Standardization. *Plastics – Determination of the Degree of Disintegration of Plastic Materials Under Simulated Composting Conditions in a Laboratory-Scale Test (ISO 20200:2023)* (International Organization for Standardization, 2023).
13. N. Hajilou, S. S. Mostafayi, A. L. Yarin, & T. Shokuhfar, A comparative review on biodegradation of poly(Lactic Acid) in soil, compost, water, and wastewater environments: Incorporating mathematical modeling perspectives. *AppliedChem*, 5(1) (2025) 1. https://doi.org/10.3390/appliedchem5010001
14. M. Fernandes, A. Salvador, M. M. Alves, & A. A. Vicente, Factors affecting polyhydroxyalkanoates biodegradation in soil. *Polymer Degradation and Stability*, 182 (2020) 109408. https://doi.org/10.1016/j.polymdegradstab.2020.109408
15. M. Barletta, C. Aversa, M. Ayyoob, A. Gisario, K. Hamad, M. Mehrpouya, & H. Vahabi, Poly(butylene succinate) (PBS): Materials, processing, and industrial applications. *Progress in Polymer Science*, 132 (2022) 101579. https://doi.org/10.1016/j.progpolymsci.2022.101579
16. W. Yu, L. Sun, M. Li, M. Li, W. Lei, & C. Wei, FDM 3D printing and properties of PBS/PLA blends. *Polymers*, 15(21) (2023) 4305. https://doi.org/10.3390/polym15214305
17. Q. Ou-Yang, B. Guo, & J. Xu, Preparation and characterization of poly(butylene succinate)/polylactide blends for fused deposition modeling 3D printing. *ACS Omega*, 3(10) (2018) 14309–14317. https://doi.org/10.1021/acsomega.8b02549
18. X. Yang, Y. Wang, Y. Zhou, J. Chen, & Q. Wan, The application of polycaprolactone in three-dimensional printing scaffolds for bone tissue engineering. *Polymers*, 13(16) (2021) 2754. https://doi.org/10.3390/polym13162754
19. S. K. Hedayati, A. H. Behravesh, S. Hasannia, A. Bagheri Saed, & B. Akhoundi, 3D printed PCL scaffold reinforced with continuous biodegradable fiber yarn: A study on mechanical and cell viability properties. *Polymer Testing*, 83 (2020) 106347. https://doi.org/10.1016/j.polymertesting.2020.106347
20. M. Fogašová, S. Figalla, L. Danišová, E. Medlenová, S. Hlaváčiková, Z. Vanovčanová, L. Omaníková, A. Baco, V. Horváth, M. Mikolajová, J. Feranc, J. Bočkaj, R. Plavec, P. Alexy, M. Repiská, R. Přikryl, S. Kontárová, A. Báreková, M. Sláviková, M. Koutný, A. Fayyazbakhsh, M. Kadlečková, PLA/PHB-based materials fully biodegradable under both industrial and home-composting conditions. *Polymers*, 14(19) (2022) 4113. https://doi.org/10.3390/polym14194113
21. N. K. Kalita, N. A. Damare, D. Hazarika, P. Bhagabati, A. Kalamdhad, & V. Katiyar, Biodegradation and characterization study of compostable PLA bioplastic containing algae biomass as potential degradation accelerator. *Environmental Challenges*, 3 (2021) 100067. https://doi.org/10.1016/j.envc.2021.100067
22. S. Momeni, K. Craplewe, M. Safder, S. Luz, D. Sauvageau, & A. Elias, Accelerating the biodegradation of poly(lactic acid) through the inclusion of plant fibers: A review of recent advances. *ACS Sustainable Chemistry & Engineering*, 11(42) (2023) 15146–15170. https://doi.org/10.1021/acssuschemeng.3c04240
23. J. O'Loughlin, H. McDonnell, R. Lawless, S. M. Kelleher, S. Fahy, B. Freeland, K. D. Rochfort, & J. Gaughran, Manufacturing process significantly impacts the rate of degradation of polylactic acid (PLA) under controlled composting conditions. *RSC Sustainability*, 4 (2026) 813–828.. https://doi.org/10.1039/D5SU00623F
24. M. Karamanlioglu, R. Preziosi, & G. D. Robson, Abiotic and biotic environmental degradation of the bioplastic polymer poly(lactic acid): A review. *Polymer Degradation and Stability*, 137 (2017) 122–130. https://doi.org/10.1016/j.polymdegradstab.2017.01.009

25. M. McMahon. (2024, March 29). Metal powders in additive manufacturing: An exploration of sustainable production, usage and recycling. *Metal-AM*. Retrieved February 8, 2026, from www.metal-am.com/articles/metal-powders-in-additive-manufacturing-an-exploration-of-sustainable-production-usage-and-recycling/
26. Höganäs AB. (n.d.). Climate and energy. Höganäs Sustainability. Retrieved February 8, 2026, from www.hoganas.com/en/sustainability/climate-and-energy/
27. H. Xiao, B. Gao, S. Yu, B. Liu, & S. Cao, Life cycle assessment of metal powder production: A Bayesian stochastic Kriging model-based autonomous estimation. *Autonomous Intelligent Systems*, 4 (2024) 20. Springer. https://doi.org/10.1007/s43684-024-00079-5
28. A. Iliaifar. (2025, June 5). Recycled metal in aerospace: Proven practice, evolving potential. Continuum Powders. Retrieved February 8, 2026, from www.continuumpowders.com/recycled-metal-in-aerospace-proven-practice-evolving-potential/
29. M. McMahon. (2024, March 29). Metal powders in additive manufacturing: An exploration of sustainable production, usage and recycling. *Metal-AM*. Retrieved February 8, 2026, from www.metal-am.com/articles/metal-powders-in-additive-manufacturing-an-exploration-of-sustainable-production-usage-and-recycling/
30. AMAZEMET. (2025, July 23). Metal powder recycling for sustainable additive manufacturing. AMAZEMET. Retrieved February 8, 2026, from www.amazemet.com/metal-powder-recycling-for-sustainable-am/
31. M. Benedetti, M. Perini, M. Vanazzi, A. Giorgini, G. Macoretta, & C. Menapace, Atomized scrap powder feedstock for sustainable Inconel 718 additive manufacturing via laser powder bed fusion: A study of static and fatigue properties. *Progress in Additive Manufacturing*, 9(6) (2023) 1843–1856. https://doi.org/10.1007/s40964-023-00547-z
32. V. Yakubov, H. Østergaard, S. Bhagavath, C. L. A. Leung, J. Hughes, E. Yasa, M. Khezri, S. K. Löschke, Q. Li, & A. M. Paradowska, Recycled aluminium feedstock in metal additive manufacturing: A state-of-the-art review. *Heliyon*, 10(5) (2024) e27243. https://doi.org/10.1016/j.heliyon.2024.e27243
33. EOS. (2024). The benefits of recycling in additive manufacturing. Retrieved February 8, 2026, from www.eos.info/content/blog/2024/the-benefits-of-recycling-in-additive-manufacturing
34. N. Maqsood, M. Rimaškas, M. Ghobakhloo, G. Mordas, & K. Skotnicová, Additive manufacturing of continuous carbon fiber reinforced polymer composites using the material extrusion process: A review. *Advanced Composites and Hybrid Materials*, 7 (2024) 202. https://doi.org/10.1007/s42114-024-01035-w
35. A. Le Duigou, A. Barbé, E. Guillou, & M. Castro, 3D printing of continuous flax fibre reinforced biocomposites for structural applications. *Materials & Design*, 180 (2019) 107884. https://doi.org/10.1016/j.matdes.2019.107884
36. M. D. H. Beg, K. L. Pickering, J. O. Akindoyo, & C. Gauss, Recyclable hemp hurd fibre-reinforced PLA composites for 3D printing. *Journal of Materials Research and Technology*, 33 (2024) 4439–4447. https://doi.org/10.1016/j.jmrt.2024.10.082
37. G. Oliveux, D. Dandy, & G. Leeke, Current status of recycling of fibre reinforced polymers: Review of technologies, reuse and resulting properties. *Composites Part A: Applied Science and Manufacturing*, 83 (2015) 7–16. https://doi.org/10.1016/j.compositesa.2015.07.008
38. D. Hidalgo-Carvajal, Á. H. Muñoz, J. J. Garrido-González, R. Carrasco-Gallego, & V. Alcázar-Montero, Recycled PLA for 3D printing: A comparison of recycled PLA filaments from waste of different origins after repeated cycles of extrusion. *Polymers*, 15(17) (2023) 3651. https://doi.org/10.3390/polym15173651

39. R. Djonyabe Habiba, C. Malça, & R. Branco, Exploring the potential of recycled polymers for 3D printing applications: *A Review. Materials*, 17(12) (2024) 2915. https://doi.org/10.3390/ma17122915
40. M. Hassan, A. K. Mohanty, & M. Misra, 3D printing in upcycling plastic and biomass waste to sustainable polymer blends and composites: A review. *Materials & Design*, 237 (2024) 112558. https://doi.org/10.1016/j.matdes.2023.112558
41. 3Devo. (2024). Sustainable 3D printing: Create and use recycled filament. Retrieved February 8, 2026, from www.3devo.com/blog/sustainable-3d-printing-create-use-recycled-filament
42. BigRep. (2024). Bio-based and recycled 3D printer filaments. Retrieved February 8, 2026, from https://bigrep.com/posts/bio-based-and-recycled-3d-printer-filaments/.
43. B. Freitas, V. Richhariya, M. Silva, A. Vaz, S. F. Lopes, & Ó. Carvalho, A review of hybrid manufacturing: Integrating subtractive and additive manufacturing. *Materials*, 18(18) (2025) 4249. https://doi.org/10.3390/ma18184249
44. L. Ben Said, B. Ayadi, S. Alharbi, & F. Dammak, Recent advances in additive manufacturing: A review of current developments and future directions. *Machines*, 13(9) (2025) 813. https://doi.org/10.3390/machines13090813
45. A. F. Ghasemi, & J. P. Duarte, A systematic review of innovative advances in multi-material additive manufacturing: Implications for architecture and construction. *Materials*, 18(8) (2025) 1820. https://doi.org/10.3390/ma18081820
46. Stratasys. (2026). J750–J735 PolyJet specification sheet (Version 1.0) [PDF]. Stratasys. Retrieved February 8, 2026, from https://support.stratasys.com/en/Printers/PolyJet-Legacy/J735-J750
47. A. Isiani, K. Crittenden, L. Weiss, O. Odirachukwu, R. Jha, O. Johnson, & O. Abika, In-situ monitoring and process control in material extrusion additive manufacturing: A comprehensive review. *Journal of Experimental and Theoretical Analyses*, 3(3) (2025) 21. https://doi.org/10.3390/jeta3030021
48. National Aeronautics and Space Administration. (n.d.). Interim, in situ additive manufacturing inspection (Reference No. MFS-TOPS-70). NASA Technology Transfer Portal. Retrieved February 8, 2026, from https://technology.nasa.gov/patent/MFS-TOPS-70
49. National Institute of Standards and Technology. (n.d.). High-resolution melt pool thermal imaging in metals additive manufacturing using two-color pyrometry (NIST Publication). Retrieved February 8, 2026, from www.nist.gov/publications/high-resolution-melt-pool-thermal-imaging-metals-additive-manufacturing-using-two-color
50. H. Li, In-situ monitoring and control in metal additive manufacturing. *Engineering Research Express*, 7(4) (2025) 040201. https://doi.org/10.1088/2631-8695/ae207e
51. A. Isiani, K. Crittenden, L. Weiss, O. Odirachukwu, R. Jha, O. Johnson, & O. Abika, In-situ monitoring and process control in material extrusion additive manufacturing: A comprehensive review. *Journal of Experimental and Theoretical Analyses*, 3(3) (2025) 21. https://doi.org/10.3390/jeta3030021
52. R. Wang, B. Standfield, C. Dou, A. C. Law, & Z. J. Kong, Real-time process monitoring and closed-loop control on laser power via a customized laser powder bed fusion platform. *Additive Manufacturing*, 66 (2023) 103449. https://doi.org/10.1016/j.addma.2023.103449
53. D. R. Gunasegaram, A. S. Barnard, M. J. Matthews, B. H. Jared, A. M. Andreaco, K. Bartsch, & A. B. Murphy, Machine learning-assisted in-situ adaptive strategies for the control of defects and anomalies in metal additive manufacturing. *Additive Manufacturing*, 81 (2024) 104013. https://doi.org/10.1016/j.addma.2024.104013

54. Sciaky. (n.d.). IRISS closed-loop control. Retrieved February 8, 2026, from www.sciaky.com/additive-manufacturing/iriss-closed-loop-control
55. Y. Cai, J. Xiong, H. Chen, & G. Zhang, A review of in-situ monitoring and process control system in metal-based laser additive manufacturing. *Journal of Manufacturing Systems*, 70 (2023) 309–326. https://doi.org/10.1016/j.jmsy.2023.07.018

6 Case Studies, Examples, and Highlights

6.1 INTRODUCTION

This chapter examines how innovation in product development, material choice, and manufacturing processes can play a central role in sustainability across industries. The primary objective is to implement and use practices that benefit the environment and society. The chapter presents real-life examples from different industries within sustainability matrices. It highlights the role of sustainability in product design, additive manufacturing, and Industry 4.0 technologies. Success stories from industries such as automobiles, aircraft, construction, consumer goods, and energy are included. Furthermore, it explores how the Industrial Internet of Things (IIoT) and digital twins enable real-time sustainability monitoring across industries. Selected example findings illustrate the purposes deliberated in earlier chapters, detailing implementation strategies, achieved outcomes, and quantifiable impacts. The emphasis is on eco-friendly practices within routine operations to achieve long-term net-zero carbon across industries [1].

6.2 EXAMPLES OF OPERATIONAL SUSTAINABILITY IN ACTION

Sustainability within industries is achieved when their day-to-day decisions aim to protect the environment, benefit communities, and ensure long-term profitability [2]. Typical measures include reducing environmental impact and waste generation, lowering energy usage, responsibly utilising materials, and upholding fair labour practices. Around the world, industries are adopting carbon-neutral technologies in their operations because of regulations and the benefits they offer [3]. The following examples show that industries can be successful in the long run if they fully follow the principles of sustainability.

Sustainable Resource Efficiency: The main objective of operational sustainability is to use resources more effectively, thereby making operations sustainable. For example, Electrolux's appliance factory in Kinston, North Carolina, implemented the 'Green Spirit' programme, which included operating production lines only as needed, repairing air leakages, and connecting sensor-activated lighting. These steps resulted in a significant drop in energy use [4].

DOI: 10.1201/9781003633754-6

Similarly, a structural composite manufacturer applied lean principles and value stream mapping to cut down on wastage in production, resulting in a 65% reduction in production costs, increased throughput from 20 to 45 units per shift, a 73% optimisation of the factory floor space, and a decrease in scrap rate from 24% to 1.8%. These data indicate that such an improvement increases efficiency, leading to economic growth and ecological advantages [5, 6].

Reduction in Waste Generation and Economic Circularity: Several industries have implemented plans to turn waste into useful materials. In the automotive industry, *Guardian Automotive* established a full recycling programme for *glass and plastics,* while Ligonier's *plant in the United States* recycled over 13,000 tons of waste in a single year, resulting in savings exceeding $360,000 [7, 8]. In aerospace, Boeing collaborates with ELG *Carbon Fibre* to recycle up to 2 million pounds of carbon fibre scraps annually, aiming to cut waste to below 20% by 2030. These examples show how important it is for manufacturing to be sustainable. In the consumer goods sector, *Unilever* obtained zero waste to landfill at more than 240 factories by 2015, five years ahead of schedule, through the best practices. These accomplishments reduced emissions by 140,000 tonnes per year, generated savings of over €220 million, and created 1,000 jobs related to sustainability. Unilever's experience demonstrates that reducing waste is important for cost savings and creating more employment opportunities [9, 10].

Renewable Energy Integration: Switching to renewable energy in manufacturing facilities is an example of operational sustainability. Automotive manufacturers are increasingly utilising renewable energy at their plants. For example, *Ford's Cologne plant* in Germany has operated exclusively on renewable electricity since 2008, sourced from local hydro and solar providers. In China, Ford's Hangzhou factory began using solar systems in 2019, now meeting 25% of its electricity demand and cutting CO_2 emissions by approximately 10,000 tonnes annually. All Audi AG units in Ingolstadt, Germany, use only green electricity, achieving carbon-neutral vehicle production, while Audi's Brussels plant became the world's first certified CO_2-neutral auto factory in 2018 [11]. In India, BMW's Chennai facility and Maruti-Suzuki plants employ on-site solar power and green power purchase agreements. These projects use renewable resources to lower costs, reduce emissions, and protect against the unpredictability of fossil fuel prices. For example, BMW's Spartanburg plant in South Carolina uses landfill methane for about 20% of its energy, reducing CO_2 emissions by 9,200 tonnes per year and generating cost savings. They show that the clean energy agreement addresses sustainability [12].

Water Conservation and Zero Liquid Discharge: Water conservation is another pillar of operational sustainability, especially in water-stressed areas. Hindustan Unilever Limited (HUL) in India runs almost all of its factories as zero liquid discharge (ZLD) sites. By the financial year of 2024–2025, 25 out of 27 HUL manufacturing sites used closed-loop water systems, treating and recycling all wastewater on-site. This shows that even water-intensive fast-moving consumer goods (FMCG) plants can greatly reduce freshwater use by reusing all process water. This is important in India, where about 40% of districts face water shortages. ITC Limited in India has maintained a water-positive approach throughout its business, including hotels, paper, and consumer goods, by putting more water back than it uses for over 23 years and staying carbon-positive for more than 20 years. Their results show that organisations are able to maintain sustainability throughout the long term. Heavy industries are also establishing ambitious water and emissions goals. UltraTech Cement, in India, aims to cut Scope 1 CO_2 intensity by 27% and Scope 2 emissions by 69% by 2032 (compared to 2017), showing a strong commitment aiming at cleaner production in high-impact sectors [13, 14].

Additive Manufacturing for Sustainability: Additive manufacturing, also known as 3D printing, serves as a tool for sustainability. They help reduce feedstock waste and allow the creation of lighter, custom products. The aerospace and automotive industries use AM to make parts and reduce scrap. Here, for example, General Electric Aviation's 3D-printed fuel nozzle combines over 20 subcomponents into a single unit, resulting in a lighter, more durable part and boosting fuel efficiency while reducing emissions over its lifetime [15].

AM is naturally more waste-efficient; a recent research study shows that improving the Multi Jet Fusion (MJF) printing process with sustainable value stream mapping reduced both material scrap and energy use per part. Flex Limited's Sorocaba site in Brazil, recognised as a World Economic Forum Sustainability Lighthouse, uses AM to 3D-print spare parts for refurbished electronics, which not only increases the product life cycle but also reduces raw material requirements and cuts waste. By using IoT and smart utilities, the facility also achieved major emission reductions. Companies that make sustainability a key part of their business have an edge over their competitors. Formulating clear goals, such as zero waste, 100% renewable energy, and water positivity, reduces ecological impact and improves financial performance. These targets help motivate and guide progress towards a sustainable future. This indicates that sustainable operations and business success are linked together, helping protect against legal changes and market shifts and supporting first-mover advantages [4, 16].

6.3 INDUSTRY 4.0 AND SUSTAINABLE DEVELOPMENT

Industry 4.0, encompassing IoT, robotics and automation, blockchain, big data analytics, machine learning and artificial intelligence, digital twins, and 3D printing, functions as a major enabler of sustainable development within industry. Smart manufacturing enables production systems to be linked, data-driven, and effective, thereby boosting resource efficiency and ecological protection. Digital transformation helps companies work more efficiently. It reduces energy use and waste, lowering environmental impact. This aligns with the United Nations Sustainable Development Goals, like efficient resource use, climate action, and clean energy. Industry 4.0 advances sustainability through immediate tracking and process optimisation. The Internet of Things, sensors, and devices continuously track energy use, emissions, and machine health. By analysing sensor data, organisations gain insight into performance and can identify areas for improvement. Input from physical systems helps industries better understand their processes and identify effective ways to improve them quickly. There is an intersection of predictive maintenance analytics, which can schedule equipment servicing at the required intervals, avoiding unnecessary breakdowns and eliminating unnecessary maintenance. A Deloitte study found that effective predictive analytics can yield up to 10% savings, a 20% increase in equipment uptime, and a 50% reduction in maintenance time, resulting in major energy and material savings over time. It supports the United Nations Sustainable Development Goals related to industry innovation and infrastructure towards productivity. In India, many large manufacturers are adopting IoT-based energy systems. For instance, JK Tyre's plant in Chennai used an Industrial IoT system to keep an eye on energy use, which made it possible to find problems in real time. Over several years, the plant reduced specific energy consumption by nearly 50% through digital monitoring and controls, as recognised by the Confederation of Indian Industry's GreenCo award [17]. Automation and robotics in Industry 4.0 play a crucial role in promoting sustainability initiatives. Advanced robots perform activities with high accuracy and efficiency, reducing waste and boosting overall productivity. These technologies reduce defect formation rates, resulting in less scrap, and optimise raw material usage. Additionally, collaborative robots (cobots) and automated guided vehicles improve logistics by cutting travel distances and enabling just-in-time delivery, which conserves energy. However, automation must be carried out prudently to promote social sustainability, enhance workforce skills, and create safer workplaces, consistent with the United Nations Sustainable Development Goals. Many organisations view Industry 4.0 investments as opportunities to improve worker safety and ergonomics by removing employees from hazardous or repetitive tasks and training them for higher-value supervisory roles. This approach tackles the 'equity' dimension of sustainable development, making sure that technology not only reduces costs and emissions but also improves job quality and inclusiveness [18].

Data analytics and artificial intelligence (AI) are driving system-wide gains in efficiency. Analysis of big data from production lines is able to identify areas with high energy consumption and opportunities for production and energy optimisation. Machine learning models could optimise complex processes, such as metal production or machining, to operate at settings that minimise fuel use and emissions. For example, Siemens implemented AI-driven control at its electronic manufacturing plant in Amberg, Germany, achieving self-optimising production that lowers error rates to near zero and lowers energy consumption per unit. This accomplishment contributed to the site being recognised as a Sustainability Lighthouse by the World Economic Forum. The World Economic Forum's Global Lighthouse Network demonstrates that Fourth Industrial Revolution technologies can increase energy efficiency by up to 50% and reduce emissions and other sustainability metrics. The forum marked a few factories as 'sustainability lighthouse' for their environmental achievements through Industry 4.0 technologies. By 2023, out of 150 lighthouse factories, only 17 factories received sustainability distinctions, including Schneider Electric's facilities in Hyderabad, India, and Unilever's digitised plants. These examples act as a model for scaling up across the industries for integrating digital and green innovation. A cloud-connected manufacturing execution system and Internet of Things sensors are used together by Schneider Electric's Hyderabad facility to optimise its compressed air systems and chillers.

Over four years, the Hyderabad plant reduced energy consumption by 59%, CO_2 emissions by 61%, water use by 57%, and waste recycling by 64%. These are significant figures for an operational plant. The plant operators focused on the Industry 4.0 tools on major resources, like implementing an 'Equaliser 4.0' device to intelligently manage compressors and a closed-loop chiller control system.

This approach highlights an important environmental advantage along with a more effective and robust operation, showing the many benefits of combining sustainable and digital initiatives. Schneider's manufacturing facility functions as a great example that investing in Fourth Industrial Revolution technologies can bring both ecological sustainability improvements and operational gains, such as lowering energy costs [17–19].

At Flex Ltd.'s Sorocaba facility in Brazil, the company implemented robotics, automation, IIoT, and a comprehensive digital platform to build a circular economy ecosystem for electronic waste, emphasising its devotion to sustainability. Material streams can be tracked using sensor networks, and a cloud-based system manages recycling processes, including the collection and sorting of e-waste. This digital plan led to a 94% reduction in on-site waste and a 38% reduction in plastic procurement costs. Moreover, by implementing efficient utility management, the Sorocaba plant achieved a 41% reduction in Scope 1 and 2 greenhouse gas emissions as well as a 30% reduction in water use after digitalisation. Notably, the site employs additive manufacturing to produce spare parts from recycled plastics, exemplifying how Industry 4.0 enables circular practices. For its achievements, the World Economic Forum recognised Sorocaba as the first Sustainability Lighthouse in Latin America. The case illustrates that linking systems through IoT and using data-driven decision tools can uncover new sustainability opportunities, such as recognising value in waste streams that traditional operations may miss [20, 21].

Industry 4.0 also promotes sustainable development in supply chains outside individual factories. Technologies, including blockchain and digital track-and-trace, can increase clarity about where materials come from by verifying their sources or tracking their carbon footprint throughout the supply chain. Automation in warehousing and logistics, such as AI-optimised delivery routes and autonomous electric forklifts, reduces fuel consumption and can work with smart grid systems to charge during off-peak periods when renewable energy is available [22].

In the domains of agriculture and construction, the Internet of Things (IoT) and robotics serve to optimise resource utilisation: intelligent irrigation systems that diminish water wastage and autonomous construction machinery that minimises material overruns, respectively, consequently supporting sustainable industry and infrastructure construction. Crucially, Industry 4.0's impact on sustainability isn't automatic; it depends on how technologies are implemented. Research indicates that digital kits can greatly support environmental objectives, but only if companies form clear targets and include sustainability metrics in their digital systems. For example, simply automating a process without rethinking resource use might yield efficiency gains but not optimal sustainability gains. Firms leading in this zone should adopt 'Sustainability 4.0' mindsets and use Industry 4.0 tools explicitly to drive sustainability outcomes. By linking energy analytics to emission dashboards and corporate targets, the payoffs can be substantial. One analysis found that digital twins are virtual models of operations, enabling companies to simulate process changes and, for companies such as LG Electronics, yield a 30% reduction in energy consumption and, for Procter & Gamble, a 30% reduction in inventory by optimising production planning. Both outcomes reduce waste and emissions while also saving costs. By using these techniques, industries can improve sustainability by separating production from resource consumption [23, 24].

From a development perspective, countries, including India, are utilising Industry 4.0 as an opportunity to 'leapfrog' to cleaner industrialisation. Initiatives like 'Make in India' now focus on smart and green manufacturing, and government policies promote the adoption of emerging tools and technologies for small and medium enterprises to improve both competitiveness and environmental performance. The integration of sustainability and digitalisation is seen as a way to meet India's commitments under the Paris Agreement without hindering economic growth [25].

One case study of an Indian bicycle manufacturer found that employing Industry 4.0 solutions, such as IoT-based machine monitoring and solar-powered factory operations, enabled the company to boost output while reducing carbon footprint through smart tech adoption. For example, Europe is the nation where Industry 5.0 is emerging, explicitly building on Industry 4.0 by adding a human-centric and sustainable focus, ensuring that technology contributes to the environment and society.

To summarise, Industry 4.0 technologies serve as catalysts for ecological development by enhancing operational visibility. These benefit companies that monitor their ecological effects in real time and develop techniques that balance efficiency and effectiveness in sustainability. Through aligning digital transformation efforts with sustainability goals, industries can accomplish two key aims: improving competitiveness and furthering global sustainable development goals. As shown in this case, studies indicate that when state-of-the-art technologies are implemented with sustainability as a priority, the outcomes can be transformative, reducing energy, water, and waste metrics by double digits and creating new models that guide industries towards a more sustainable future.

6.4 REAL-WORLD APPLICATIONS AND SUCCESS STORIES

To strengthen the illustration of operational sustainability principles, the sub-section highlights practical cases across industries, emphasising verifiable results and practices. These examples, drawn from both global and Indian contexts, show that industries are moving towards large-scale sustainable operations. These examples show unique approaches, from leading-edge technology adoption to clever process re-engineering, and the concrete benefits achieved in practice.

Automotive Industry: The automobile industry is under extreme pressure to decarbonise. They are not only focusing on their products but also on their manufacturing processes. Their efforts have led to some compelling success stories. One case that really illustrates the point is BMW's Leipzig plant in Germany, which has integrated on-site wind power generation to supply its production. Four large wind turbines, totalling 10 MW, on the Leipzig factory grounds produce about 21.9 GWh of electricity annually for the plant, directly meeting energy needs with zero-carbon power. This has made the plant a symbol of eco-friendly car-making. It not only makes BMW's electric cars but also uses renewable energy. This helps to lower carbon emissions during the entire life of the product. On another front, BMW's Rosslyn plant in South Africa pioneered the use of biogas energy in auto manufacturing. Partnering with a local waste-to-energy company, the plant sources approximately 30% of its energy from landfill methane gas. Over nearly two decades, this has prevented over 9,200 tonnes of CO_2 emissions each year, comparable to removing thousands of cars from the road. Moreover, continuing upgrades and a new contract aim to cut an additional 74,000 tonnes of CO_2 over eight years. The success at Rosslyn, using a readily available local waste resource to power high-tech manufacturing, is a blueprint for circular energy solutions in industry [15–18].

In India, Mahindra & Mahindra has become a champion of sustainable practices. It became the first Indian company to launch an internal carbon pricing programme, initially set at $10 per ton of CO_2, which was later increased and is now applied to evaluate every capital project. This has steered investment towards sustainable technologies, for example, solar rooftops and energy-saving equipment, by making the cost of carbon explicit in decision-making. As a result, several Mahindra auto plants achieved reductions in energy intensity and CO_2 emissions.

One of Mahindra's SUV manufacturing plants achieved 'water-positive' status by harvesting more water than it consumes, and another became zero waste to landfill, illustrating the company's efforts in energy, water, and waste minimisation. These efforts ended in Mahindra receiving India's first CII GreenCo Platinum rating in the automotive sector, a recognition of excellence in environmental performance. The main takeaway from Mahindra's story is how corporate policy innovation, like strict carbon targets, can trigger operational changes that result in sustainability achievements [26, 27].

Aerospace and Defence: Aerospace industries' production faces unique sustainability challenges. These include energy-intensive processes, exotic materials, and strict safety requirements. Still, major businesses have made significant progress. Pratt & Whitney, a leading jet engine manufacturer, has set comprehensive 2025 eco-friendly targets, supported by $60 million in funding for more than 800 projects. By 2025, it aimed for zero waste to landfill in factories, an 80% reduction in greenhouse gas emissions relative to the baseline, and an 80% cut in water usage, among other targets. Pratt & Whitney has already achieved a 30% reduction in greenhouse gas emissions at its factories and is on track to meet its 80% goal. One of its plants implemented a closed-loop cooling water system that saves millions of gallons of water each year, and several sites transitioned to renewable energy contracts. Such results show that even manufacturing processes can dramatically shrink their footprint via target-setting, investment, and innovation [4, 28].

Airbus, on the other hand, has extended sustainability beyond its own facilities to the entire aircraft life cycle. Airbus was the first aerospace original equipment manufacturer to achieve ISO 14001 certification for all production sites and products since 2007. Through its 'High5+' goals, Airbus aims to reduce manufacturing waste by 20% by 2030 and achieve zero waste to landfill, with all waste either recycled or energy recovered. Airbus also co-founded Tarmac Aerosave, a company that dismantles and recycles aircraft at the end of life, recovering up to 92% of an aircraft's weight for reuse and recycling. It is a breakthrough in addressing the sustainability of aerospace products, which historically have not been recycled. A real-world highlight is Airbus' use of additive manufacturing for aircraft parts, such as 3D-printed cabin brackets and airframe components, which reduce material usage by 30–55% compared to machined parts and cut weight, thereby boosting fuel efficiency in operation. The success story here is the holistic approach. Airbus tackled the sustainability issue through design (lightweight, recyclable planes), manufacturing (lean, waste-minimised plants), and end-of-life (a circular economy for aircraft). This overall approach considerably minimises aerospace's environmental impact and acts as a model for high-value manufacturing industries [18, 28].

Construction and Infrastructure: The construction industry is a significant consumer of raw building materials and a major source of emissions and waste. However, green building case studies prove that sustainable construction is feasible and favourable. A shining example in India is the Indira Paryavaran Bhawan in New Delhi, which serves as the headquarters of the Ministry of Environment. This building was designed to be net-zero energy and features several green structures. Through employing passive architectural design, such as optimal orientation, daylighting, and natural ventilation, and efficient active systems, such as chilled-beam air conditioning, light-emitting diode (LED) lighting with sensors, and on-site solar panels, the building consumes 70% less energy than a conventional office of its size. It also incorporates wastewater recycling and rainwater harvesting, achieving significant water conservation savings. Indira Paryavaran Bhawan earned a GRIHA 5-Star rating and a Leadership in Energy and Environmental Design (LEED) Platinum rating, demonstrating that even government infrastructure can advance sustainability. Its successful operation has inspired the adoption of green building codes in other public projects across India.

Globally, The Edge in Amsterdam, often called the world's greenest office building, employs IoT-connected lighting and heating, ventilation, and air conditioning (HVAC) systems that adapt based on occupancy, resulting in an annual energy consumption of only ~70 kWh/m², primarily supplied by renewables. Meanwhile, Singapore's Changi Airport carried out a major upgrade of its HVAC systems, installing digital controls and thermal energy storage, resulting in 40% savings in cooling energy. The Surat Metro project in India provides a success story within sustainable infrastructure development, with construction materials that include a high percentage of recycled content, and the stations are being built with solar rooftops and water harvesting, aiming for high-level green certifications. The recurring theme is that including sustainability from the design phase, whether it's a building or infrastructure, produces significant, prolonged benefits, such as reduced operating costs, greater resilience (e.g. less dependency on the water grid), and improved resident well-being. Many such projects report only marginally higher initial costs (often <5%), which are recovered through energy and water savings [18].

Consumer Goods and FMCG: In consumer goods manufacturing, sustainability efforts typically concentrate on economical resource use and re-evaluating the product life cycle. Unilever's global 'Sustainable Living Plan' is a well-documented success, particularly its manufacturing pillar, which, as previously noted, achieves zero waste to landfill. By 2020, Unilever also cut CO_2 emissions from energy use in manufacturing by 65% per ton of production, relative to a 2008 baseline, through energy efficiency and switching to renewables. A concrete example is Unilever's personal care product factory in Maharashtra, India, which implemented a biomass boiler using agri-waste as fuel to replace furnace oil, reducing that site's carbon emissions by over 5,000 tonnes/year.

Procter & Gamble (P&G) similarly launched its 'Ambition 2030' goals, one of its baby care product plants in the US achieved 100% renewable electricity and zero manufacturing waste while also reducing water use per unit by 25%. That site pioneered closed-loop recycling of diaper trimmings, feeding them back into production, saving $1 million annually in raw material costs. The success factor here is how environmental responsibility and productivity align when waste is treated as a resource and renewable energy is embraced. Consumer goods plants not only become greener but also often leaner in their cost structures.

A unique success story in consumer goods is ITC's Paperboards Unit in Bhadrachalam, India. This pulp and paper mill confronted challenges of high water and energy usage. Through a series of sustainability projects, including installing an oxygen delignification system, upgrading to energy-saving motors, and executing comprehensive water recycling, the mill reduced water consumption per tonne of paper by 60% over a decade and is now a zero effluent discharge facility. It also generates over 50% of its power from biomass and black liquor (a pulp byproduct), considerably reducing fossil fuel use. These measures caused the site to receive the CII Environmental Best Practices Award, evidencing that even traditionally high-polluting industries can reinvent themselves with clean technology and process innovation [29, 30].

Energy Sector: Energy companies are displaying their commitment by adopting more eco-friendly practices and operations. Renewable energy companies, of course, inherently drive sustainability, but even fossil-fuel-based utilities are finding ways to operate more sustainably. One example is Tata Power's Trombay thermal power station, which invested in a digital efficiency refinement system. By using AI to analyse boiler performance and IoT data from turbines, the plant improved its heat rate (thermal efficiency) by ~1.5% and reduced coal consumption; accordingly, a seemingly small percentage translates into thousands of tons of coal and CO_2 saved annually.

ONGC is India's Oil & Natural Gas Corp. It upgraded its offshore platforms with waste heat recovery units. It also optimised pumping schedules using digital twins. These changes saved 14% of energy in platform operations. Meanwhile, NextEra Energy in the US, a leading renewables operator, uses advanced predictive analytics on its wind turbine fleet (digital twins of turbines) to schedule pre-emptive maintenance and position turbine blades optimally based on weather forecasts. This has increased its wind farms' energy output by a few percentage points and extended the life of turbine components, reducing material waste from replacements. These cases demonstrate that digitalisation in the energy sector not merely increases renewable generation but also optimises its efficiency and cleanliness [31].

Finally, success stories regularly involve multi-stakeholder collaboration. The Sustainable Apparel Coalition (an alliance of global fashion brands and manufacturers) developed the Higg Index tools to measure sustainability in textile factories. In a pilot project, a denim factory located in Gurgaon, India, implemented Higg recommendations to update its processes: substituting chemical-heavy washing techniques with enzyme-based methods, installing highly efficient dyeing machines, and powering the facility with biomass briquettes.

The result was a 50% reduction in water use, 30% energy savings per pair of jeans, and improved effluent quality. The project's success was due to effective knowledge sharing within the coalition and clearly defined metrics. It highlights how industry-wide initiatives shall advance sustainability by promoting best practices and establishing common standards [31].

There are common themes in these different stories, including devotion to leadership, creativity, and exacting measurement. The companies achieved technological advances by setting targets that often exceeded regulatory requirements and by focusing on performance data to promote continuous improvement. The quantifiable results, such as tens of thousands of tonnes of CO_2 avoided, significant percentage reductions in resource consumption, or notable cost savings, make a strong case that sustainable operations are practical and beneficial. These success stories often mention additional benefits, such as improved reputation, easier regulatory compliance, and greater employee pride, as people prefer to work for environmentally responsible organisations. As more case studies like these emerge, they serve as inspiration and proof of concept for other companies to strengthen their operational-level sustainability efforts, creating a positive cycle of learning and progress across industries [32–34].

6.5 IOT AND DIGITAL TWINS FOR REAL-TIME SUSTAINABILITY MONITORING

An essential aspect of attaining operational sustainability is the instant monitoring and management of sustainability performance. This is where the Internet of Things (IoT) and digital twin technology have become game changers. Deploying networks of sensors, known as the IoT, creates virtual replicas of physical processes. This allows safe and easy interaction with physical properties. The technology becomes increasingly accessible and engaging. Digital twins enable organisations to continuously monitor resource use and environmental parameters. They can also predict outcomes and improve processes instantly. The part discusses how IoT and digital twins help monitor sustainability metrics such as energy, emissions, and resource efficiency, and how they keep operations green.

IoT for Real-time Supervision: IoT devices, such as smart metres, temperature and humidity sensors, equipment vibration sensors, and emission monitors, are now universal in modern facilities. These sensors feed data to cloud-based platforms or on-site control systems, providing visibility into every facet of operations [35, 36]. To support sustainability, it is important to have immediate awareness of any defects or inefficiencies so they can be addressed them quickly and effectively. A sudden increase in electricity usage or a drop in power factor on a certain production line can be immediately detected by an IoT-based energy management system, triggering an alert to facility managers. Rockwell Automation reports that combining data from multiple physical systems and sensors helps clients identify energy usage and emission sources in real time, effectively establishing an 'operational sustainability dashboard' for production managers [37].

One example is Althofen, Austria, a smart factory site operated by Flex. The site achieved carbon-neutral status, in part, by using IoT energy monitors to monitor real-time power consumption and detect leaks and losses. The system alerts staff to issues such as compressed air leaks or equipment left running, which can then be fixed promptly. This IoT-enabled vigilance helped Althofen avoid energy inefficiency and time-consuming operations, such as equipment charging, during off-peak renewable energy availability. Additionally, the site's environmental data, such as energy and water use, is displayed live on a CO_2 dashboard that both managers and customers can access, fostering openness and responsibility. Many companies now utilise equivalent IoT-driven dashboards for sustainability KPIs, monitoring hourly energy use per unit produced, water use per batch, real-time emissions levels, etc., to ensure they stay within targets. This prompt feedback system is important for achieving ambitious sustainable targets and quickly correcting course if metrics drift [38].

Digital Twins for Simulation and Optimisation: A digital twin is a virtual model of a physical asset or process that reproduces its state in real time using sensor data. But beyond passive mirroring, digital twins enable scenario simulation and predictive analytics. This is tremendously useful for sustainability. With a digital twin of a factory, for instance, engineers can test adjustments in the virtual world, such as changing a machine's operating schedule or altering HVAC settings in a building, to see how it would impact energy use, without risking disruptions in the real world. They can also simulate rare events, such as power outages and extreme weather, to improve resilience along with emergency response competencies [39, 40].

A leading example is Foxconn's use of a digital twin for a new electronics factory in Mexico. Foxconn built a highly detailed virtual plant using NVIDIA's Omniverse platform that replicates production lines, robotic systems, and utility infrastructure. As a result, Foxconn's engineers could optimise the layout for energy and material flow even before the physical factory was completed. They tested robot placements and sensor networks in the digital twin to maximise automation efficiency and limit human intervention in unsafe areas. The Foxconn twin also enabled them to execute virtual training for AI vision systems and robotics. Importantly, Foxconn's chairman reported that this approach is predicted to reduce the factory's energy consumption by over 30% annually, based on improvements identified in simulations. This is a staggering figure, translating not only into cost savings but also into a huge cut in the carbon footprint for such a large plant. It shows how digital twins can help organisations be more proactive about sustainability. For example, they can help us fix problems in the model, such as removing bottlenecks that waste energy before they become real energy drains. Foxconn's success shows that digital twin analysis can help any complex facility operate better by more than 10%. This is a great chance for long-term growth.

Digital twins are also being used for building energy management. For instance, a digital twin of an office building in the Netherlands was integrated with its building management system. By simulating occupancy patterns and HVAC responses, the building operators fine-tuned the control algorithms, yielding a comfortable environment while minimising energy. This contributed to the building becoming energy-positive, generating more energy than it consumes. Likewise, entire city districts are getting digital twins to improve utilities, as in a pilot in France where a digital twin helped coordinate between solar generation, battery storage, and building demands, improving renewable utilisation and cutting peak grid load. These cases illustrate that digital twins can operate at multiple scales, such as equipment, factory, building, and grid, to find sustainability gains [39].

Integration of IoT and Digital Twins: IoT supplies the real-time information stream, while digital twins offer the analytical as well as predictive engine; together, they form an effective closed-loop system for ongoing improvement. For example, Siemens Energy implemented digital twins of turbines, coupled with IoT sensor data, to continuously adjust combustion parameters, elevating efficiency and reducing emissions (NO_x, CO_x) under shifting loads. In process industries such as chemicals or food processing, digital twins of the process, trained on historical IoT data, can recommend optimal setpoints to operators that reduce energy or water use while maintaining throughput. These recommendations can be updated in real time as conditions change. Companies like Procter & Gamble have implemented digital twins for entire production lines; one result was identifying ways to reduce line startup time, which historically ran idle during ramp-up, and thereby save energy and materials. P&G achieved significant operational savings and decreased waste per batch after acting on twin simulations, reinforcing that coupling IoT data with digital modelling yields effective insights. Another interesting application is in supply chain sustainability. Some firms have digital twins of their logistics network, fed by IoT trackers on vehicles and inventory. They can simulate various delivery schedules or modal mixes (e.g. truck vs. rail) to reduce fuel use. This assisted a consumer goods company in reducing its distribution emissions by 15% through redesigning routes and consolidating loads virtually before physical implementation.

Real-Time Emission Control and Monitoring: IoT detectors are also essential for monitoring air and water emissions. Continuous emissions monitoring systems (CEMS) on smokestacks provide vital data to compliance systems and digital models of environmental impact, helping us better understand and protect our environment. For instance, a steel plant in Europe uses an IoT-based monitoring network around its perimeter and a digital twin of pollutant dispersion. When sensors detect elevated particulate matter, the twin helps attribute it to a specific process in the plant, permitting operators to immediately adjust that process, for example, tweaking a baghouse filter cleaning cycle. This closed loop has helped the plant stay well below regulatory limits and even facilitated early adjustments on days with adverse weather that would otherwise have trapped pollution. In water management, the deployment of IoT probes along effluent streams, in conjunction with a digital twin of the wastewater treatment plant, facilitates the prompt adjustment of treatment dosages to maintain parameters such as pH within desired ranges, thereby preventing unintentional discharge problems.

Government and Policy Effects: Policymakers acknowledge the importance of IoT and digital twins in sustainability monitoring. Some environmental regulations are moving towards calling for continuous data reporting rather than periodic audits.

Digital infrastructure in factories will make it easier for companies to comply and even earn incentives. For example, under India's Perform–Achieve–Trade (PAT) scheme for energy efficiency, companies with instant monitoring can more readily demonstrate improvements and trade energy-saving certificates. Governments are also funding digital twin testbeds; the U.S. Department of Energy announced $50 million in funding for smart manufacturing digitalisation to help micro, small, and medium-sized enterprises diminish their environmental footprint. This support recognises that digital tools are essential enablers for achieving national sustainability targets [41].

Challenges and Future Outlook: Although IoT and digital twins offer clear advantages, challenges such as data silos, integration issues, and the need for trained staff to interpret and use the data persist. Industries should focus on investing in technology, as well as training and process improvements, to fully realise the value of their efforts. Cybersecurity is a primary concern because interconnected critical systems heighten the importance of information precision and security. Even with these problems, more people are starting to use digital twins. The costs of sensors and computers have gone down, and stories of early success are making others more confident. In the future, we might see even more realistic digital twins, where virtual reality tools allow managers to walk through a factory and observe how everything operates.

This could help refine decision-making speed and sustainability. In conclusion, IoT and digital twins transform sustainability monitoring from a retrospective to an ongoing, proactive practice. Real-time information and simulations help operators make instant adjustments to keep processes within optimal and sustainable ranges. The capacity to anticipate problems and test improvements virtually accelerates sustainability innovation. Companies that adopt these tools report significant gains, such as Foxconn's 30% reduction in energy through virtual factory optimisation, Schneider's 50–60% resource savings via IoT-based controls, and LG's 30% reduction in energy through digital twin technology. These are not minor tweaks but substantial step-change improvements. As this system becomes more widespread, more industries will be able to achieve real-time sustainability, creating an operational environment where efficiency and low environmental impact are automatically maintained, akin to an autopilot for green operations. This is promising for meeting corporate eco-friendly goals and, on a larger scale, for industries worldwide committed to climate and sustainability targets. The combination of digital technology alongside sustainability thus appears as a key element of operational excellence in the 21st century, confirming that developments are lasting, clear, and in line with our environmental limits.

REFERENCES

1. B. M. Omowole, A. Q. Olufemi-Phillips, O. C. Ofodile, N. L. Eyo-Udo, & S. E. Ewim, Conceptualizing green business practices in SMEs for sustainable development. *International Journal of Management & Entrepreneurship Research*, 6 (2024) 3778. https://doi.org/10.51594/ijmer.v6i11.1719.

2. A. A. E. Abdelraheem, A. M. H. Musa, M. E. M. Khogly, & Y. Elboukhari, Consolidating sustainability efforts: The role of effective supply chain management in balancing economic growth, environmental stewardship, and social responsibility. *Journal of Project Management*, 10 (2025) 533. https://doi.org/10.5267/j.jpm.2025.4.001.
3. A. Singh, Sustainability practices in business operations. *International Journal for Research Publication and Seminars*, 15 (2024) 18. https://doi.org/10.36676/jrps.v15.i3.1424.
4. P. Saxena, P. Stavropoulos, J. D. Kechagias, & K. Salonitis, Sustainability assessment for manufacturing operations. *Energies*, 13 (2020) 2730. https://doi.org/10.3390/en13112730.
5. M. J. A. Pinto, & J. V. Mendes, Operational practices of lean manufacturing: Potentiating environmental improvements. *Journal of Industrial Engineering and Management*, 10 (2017) 550. https://doi.org/10.3926/jiem.2268.
6. J. Monteiro, F. J. G. Silva, C. R. Regattieri, L. P. Ferreira, R. Lucas, & R. C. M. Sales-Contini, Improving the economic and environmental sustainability of painting and glueing operations in a luxury leather goods industry using DMAIC methodology – a case study. *Journal of Material Cycles and Waste Management*, 27 (2025) 4264–4284. https://doi.org/10.1007/s10163-025-02357-7.
7. S. Faucheux, I. Nicolaï, & M. O. 'Connor, Economic globalisation, competitiveness and environment. *HAL (Le Centre pour la Communication Scientifique Directe)* Post-Print hal-01799300, HAL (1997).
8. N. Sharma, S. B. Ingole, H. S. Pokhariya, A. Parmar, K. C. Shilpa, U. M. Reddy, & H. A. Hussny, From waste to worth management: A comprehensive intelligent approach to resource utilization and waste minimization. *E3S Web of Conferences*, 453 (2023) 1029. https://doi.org/10.1051/e3sconf/202345301029.
9. F. Meng, Y. Cui, S. Pickering, & J. McKechnie, From aviation to aviation: Environmental and financial viability of closed-loop recycling of carbon fibre composite. *Composites Part B Engineering*, 200 (2020) 108362. https://doi.org/10.1016/j.compositesb.2020.108362.
10. A. Darmandieu, C. Garcés-Ayerbe, A. Renucci, & P. R. Torres, How does it pay to be circular in production processes? Eco-innovativeness and green jobs as moderators of a cost-efficiency advantage in European small and medium enterprises. *Business Strategy and the Environment*, 31 (2021) 1184. https://doi.org/10.1002/bse.2949.
11. T. Princz-Jakovics, N. C. Nagypál, & T. Pálvölgyi, Comprehensive assessment of the carbon neutrality actions from the automotive sector. *Environment Development and Sustainability*, 27 (2023) 23909–23928. https://doi.org/10.1007/s10668-023-04274-x.
12. E. Savran, Ö. Yavaş, B. E. Nalbur, & F. Karpat, Energy and carbon loss management in an electric bus factory for energy sustainability. *Transdisciplinary Journal of Engineering & Science*, 13 (2022) 97–110. https://doi.org/10.22545/2022/00207.
13. S. Dhar, M. Pathak, & P. R. Shukla, Transformation of India's steel and cement industry in a sustainable 1.5 °C world. *Energy Policy*, 137 (2019) 111104. https://doi.org/10.1016/j.enpol.2019.111104.
14. D. Silva, C. Prolo, P. H. T. Braga, A. Witzke, T. Gomes, & L. Lotero, Corporate emissions and climate targets: Insights from high-emitting firms in Brazil's transition to sustainability. *PLOS Sustainability and Transformation*, 4 (2025) e0000198. https://doi.org/10.1371/journal.pstr.0000198.
15. H. Hegab, N. Khanna, N. Monib, & A. Salem, Design for sustainable additive manufacturing: A review. *Sustainable Materials and Technologies*, 35 (2023) e00576. https://doi.org/10.1016/j.susmat.2023.e00576.
16. S. Polo, E. M. Rubio, J. Ayllón, & B. de Agustina, Emerging advances in sustainable manufacturing. *Processes*, 13 (2025) 1549. https://doi.org/10.3390/pr13051549.

17. A. Santanaa, P. Afonsoa, A. Zaninb, & R. Wernkeb, Costing models for capacity optimization in Industry 4.0: Trade-off between used capacity and operational efficiency. *Procedia Manufacturing*, 13 (2017) 1183–1190.
18. P. Nyamekye, R. Lakshmanan, V. Tepponen, & S. Westman, Sustainability aspects of additive manufacturing: Leveraging resource efficiency via product design optimization and laser powder bed fusion. *Heliyon*, 10(1) (2023) e23152. https://doi.org/10.1016/j.heliyon.2023.e23152.
19. S. Nazir, A. Z. Piprani, S. Mukhuty, A. Upadhyay, J. Vilko, & W. C. Poon, Industry 4.0 integration for sustainability and value creation: Moderating role of digital and environmental strategy. *Business Strategy and the Environment*, 35(2) (2025) 2192–2209. https://doi.org/10.1002/bse.70283.
20. S. Achar, Influence of IoT technology on environmental monitoring. *Asia Pacific Journal of Energy and Environment*, 7 (2020) 87. https://doi.org/10.18034/apjee.v7i2.649.
21. S. Strazzullo, L. Cricelli, C. Troise, & M. A. Camilleri, Leveraging industry 4.0 technologies for sustainable value chains: Raising awareness on digital transformation and responsible operations management. *Sustainable Development* 33(2) (2024) 2189–2202. https://doi.org/10.2139/ssrn.4964036.
22. M. Trstenjak, M. Mustapić, P. Gregurić, & T. Opetuk, Use of green industry 5.0 technologies in logistics activities. *Tehnički glasnik*, 17 (2023) 471. https://doi.org/10.31803/tg-20230518185836.
23. P. K. K. Singh, Measuring the broader value proposition of digital transformation in supply chains. *International Journal of Supply Chain Management*, 13 (2024) 16. https://doi.org/10.59160/ijscm.v13i1.6222.
24. S. Sengupta, & H. C. Dreyer, *Digitalization and Sustainability*. Routledge eBooks (Informa, 2024), p. 98. https://doi.org/10.4324/9781032693415-8.
25. K. Kolaro, G. Pitić, E. Vlačić, & U. Milosavljević, Competitiveness and sustainability in small and open economies in the age of industry 5.0. *Ekonomika preduzeca*, 71 (2023) 113. https://doi.org/10.5937/ekopre2302113k.
26. M. Moslehpour, D. Ekowati, & J. Sulistiawan, Corporate sustainability practices in Indian automobile industry: Enhancing government initiatives, economic improvements, and environmental practices. *Engineering Economics*, 34 (2023) 456. https://doi.org/10.5755/j01.ee.34.4.33160.
27. Q. He, The influence of organizational policies on firm environmental performance through sustainable technologies and innovation and stakeholder concerns. *Scientific Reports*, 15 (2025) 10019. https://doi.org/10.1038/s41598-025-94499-9.
28. M. Gopal, H. G. Lemu, & E. M. Gutema, Sustainable additive manufacturing and environmental implications: Literature review. *Sustainability*, 15 (2022) 504. https://doi.org/10.3390/su15010504.
29. R. Sims, P. Mercado, W. Krewitt, G. Bhuyan, D. Flynn, H. Holttinen, G. D. M. Jannuzzi, S. Khennas, Y. Liu, L. J. Nilsson, J. M. Ogden, K. Ogimoto, M. O'Malley, H. Outhred, Ø. Ulleberg, F. V. Hulle, M. Bazilian, M. Beerepoot, T. Demayo, E. Denny, D. Infield, A. Keane, A. Lee, M. Milligan, A. Mills, M. Power, P. D. Smith, L. Söder, A. Tuohy, F. Ueckerdt, J. Zhang, J. Skea, & K. Strunz, Integration of renewable energy into present and future energy systems. In O. Edenhofer, R. Pichs-Madruga, Y. Sokona, K. Seyboth, S. Kadner, T. Zwickel, P. Eickemeier, G. Hansen, S. Schlömer, C. von Stechow, & P. Matschoss (Eds.), *Cambridge University Press eBooks* (Cambridge University Press, 2011), p. 609. https://doi.org/10.1017/cbo9781139151153.012.
30. A. Mani, T. Budd, & E. Maine, Emissions-intensive and trade-exposed industries: Technological innovation and climate policy solutions to achieve net-zero emissions by 2050. *RSC Sustainability*, 2 (2023) 903. https://doi.org/10.1039/d3su00335c.

31. R. U. Attah, B. M. P. Garba, I. Gil-Ozoudeh, & O. Iwuanyanwu, Digital transformation in the energy sector: Comprehensive review of sustainability impacts and economic benefits. *International Journal of Advanced Economics*, 6 (2024) 760. https://doi.org/10.51594/ijae.v6i12.1751.
32. S. Yue, G. Ye, F. Khalid, & T. Wang, From digital transformation to energy efficiency: Ownership concentration's hidden role in driving climate solutions. *Humanities and Social Sciences Communications*, 12 (2025) 1–4. https://doi.org/10.1057/s41599-025-05911-8.
33. C. A. Arinze, O. A. Ajala, C. C. Okoye, O. C. Ofodile, & A. I. Daraojimba, Evaluating the integration of advanced it solutions for emission reduction in the oil and gas sector. *Engineering Science & Technology Journal*, 5 (2024) 639. https://doi.org/10.51594/estj.v5i3.862.
34. M. Kim, J. Y. Shim, S. Lim, H.-D. Lee, S. C. Kwon, S. Hong, & S.-J. Ryu, Reduction of greenhouse gas emissions by optimizing the textile dyeing process using digital twin technology. *Fashion and Textiles*, 11 (2024) 17. https://doi.org/10.1186/s40691-024-00384-w.
35. B. Ameh, Digital tools and AI: Using technology to monitor carbon emissions and waste at each stage of the supply chain, enabling real-time adjustments for sustainability improvements. *International Journal of Science and Research Archive*, 13 (2024) 2741. https://doi.org/10.30574/ijsra.2024.13.1.1995.
36. A. Badhoutiya, H. Darokar, R. P. Verma, M. Saraswat, S. Devaraj, V. H. Raj, & Z. N. Abdulhussain, Regenerative manufacturing: Crafting a sustainable future through design and production. *E3S Web of Conferences*, 453 (2023) 1038. https://doi.org/10.1051/e3sconf/202345301038.
37. A. I. Ibokette, T. O. Ogundare, E. O. Danquah, A. P. Anyebe, J. A. Agaba, & T. M. Olala, The impacts of emotional intelligence and IOT on operational efficiency in manufacturing: A cross-cultural analysis of Nigeria and the US. *Computer Science & IT Research Journal*, 5 (2024) 1978. https://doi.org/10.51594/csitrj.v5i8.1464.
38. G. Rolofs, F. Wilking, S. Goetz, & S. Wartzack, Integrating digital twins and cyber-physical systems for flexible energy management in manufacturing facilities: A conceptual framework. *Electronics*, 13 (2024) 4964. https://doi.org/10.3390/electronics13244964.
39. Y. Yakovenko, & R. Shaptala, Study of digital twins as the driving force of digital transformation and achieving the goals of sustainable development. *Technology Audit and Production Reserves*, 2 (2024) 11. https://doi.org/10.15587/2706-5448.2024.301423.
40. L. V. Bandara, & L. Buics, Digital twins in sustainable supply chains: A comprehensive review of current applications and enablers for successful adoption. *Engineering Proceedings*, 79(1) (2024) 64. https://doi.org/10.3390/engproc2024079064.
41. Md. A. Shabur, A comprehensive review on the impact of Industry 4.0 on the development of a sustainable environment. *Discover Sustainability*, 5 (2024) 97. https://doi.org/10.1007/s43621-024-00290-7.

7 Challenges and Future Directions

Operational sustainability has become an important goal in modern manufacturing systems. It also matters because manufacturing processes use large amounts of energy and materials. While earlier chapters discussed materials, design methods, additive manufacturing processes, and real-world applications with success stories, challenges remain that limit the widespread adoption of the concept [1]. These challenges are not only technical but also economic, social, and organisational. This chapter discusses the barriers that industries face today and outlines future directions to strengthen the usability of operational sustainability [2]. Sustainable manufacturing is often described as the efforts to meet economic, environmental, and social goals together [3]. Special focus is given to material scarcity and practical solutions that can support long-term sustainable manufacturing. This book focuses on three linked pillars for operational sustainability. The first pillar is materials selection, the second is product design, and the third is additive manufacturing and related process choices.

7.1 EMERGING GLOBAL CHALLENGES: MATERIAL SCARCITY

Material scarcity is now a central risk for sustainable manufacturing. It affects both clean energy systems and modern manufacturing tools [4]. Many of the materials at risk are also important for additive manufacturing feedstock and parts. This includes lithium, cobalt, nickel, graphite, copper, and rare earth elements. Demand is rising fast. Clean energy is now a main driver of demand growth for many critical minerals. There was strong growth reported in 2023. Lithium demand rose by about 30%. Demand for nickel, cobalt, graphite, and rare earth elements rose by about 8% to 15%. Government data shows how strongly some end uses now dominate. For lithium, global end use is estimated to be led by batteries at about 87%. Batteries are used in electric vehicles, devices, tools, and grid storage [5]. In 2024, global lithium production grew by about 18% to around 240,000 tonnes (lithium content). Global consumption was estimated at about 220,000 tonnes. The same pressures appear in cobalt. It was estimated that mine cobalt production reached about 290,000 tonnes in 2024 (cobalt content). It also reported a very high concentration in one producer. It accounted for an estimated 76% of world cobalt mine production in 2024. This level of concentration raises supply risk. Rare earth elements show similar patterns.

DOI: 10.1201/9781003633754-7

Global rare earth mine production was estimated at around 390,000 tonnes in 2024 (rare-earth-oxide equivalent). was estimated at around 270,000 tonnes. was estimated at around 2,900 tonnes. The same source notes that recycling is still limited, with only small quantities recovered from items such as batteries, magnets, and fluorescent lamps [6].

Scarcity is not only about mining. Processing and refining often create the tightest choke points. The IEA notes that markets have become more concentrated in refining and processing. For a group of key minerals, the average market share of the top three refining nations rose to about 86% in 2024, up from around 82% in 2020 [7].

These patterns make supply chains sensitive to politics and trade action. Export controls, tariffs, and licensing rules can change lead times and prices. The cobalt chapter in the USGS data, for example, notes that tariff changes were enacted on some cobalt-related imports in 2024. This type of policy action can affect cost and availability for downstream firms [8, 9].

Material scarcity also has direct environmental and social costs. Mining and processing can be energy-intensive. They can also cause pollution and loss of nature. Social risks include harm to local communities and serious labour rights concerns. It enlists greenhouse gas emissions from mining and processing, biodiversity loss, pollution, and human rights abuses as key concerns linked to critical minerals.

Independent studies also show that impacts vary by metal and region. A 2025 report on the social impacts of mining critical raw materials describes issues such as water stress, land conflict, and resettlement risks in different settings. This matters for operational sustainability because upstream harm can become a business risk [3, 10].

Recycling should reduce pressure on the virgin supply. Yet recycling systems are still not strong enough for many materials. Plastics are a clear example. Reports indicate that only about 9% of plastic waste is recycled. The rest is landfilled, burned, or mismanaged. This weak circular flow increases demand for virgin feedstock and keeps waste volumes high. Batteries are another key case. Recycling of battery materials can reduce new demand. But capacity and collection must grow. The IEA notes that if all announced projects are built on time, global battery recycling capacity could exceed 1,500 GWh by 2030. It also shows that planned capacity is unevenly distributed by region [11, 12].

Some materials are difficult to recycle in practice. Composites are a major challenge. Many fibre-reinforced plastics, especially thermoset composites, are hard to recycle at scale. Reviews note that recycling routes can be complex and may require costly facilities. They also note that end-of-life management remains one of the hardest sustainability issues for these materials.

The same problem appears in additive manufacturing when composite filaments and multi-material parts become common. Recycling becomes harder when a part contains mixed phases or fillers. A review focused on polymer composite printing notes that adding fillers can increase processing steps and create more disposal challenges. It also argues that recycling is essential but not always simple for such feedstock.

Recycled polymers and recycled feedstocks are promising, but they face real limits. Studies on recycled polymers for 3D printing point to common issues.

Recycled materials can vary in quality and may contain impurities. This can reduce print quality and part strength. It can also cause process problems such as poor flow and poor layer bonding [13].

Bio-based materials can reduce dependence on fossil feedstocks. Yet they also raise questions about land use, food security, and nature. A 2024 assessment of the bioeconomy notes that demand for bio-based products and fuels can compete with food production and can change land-use patterns [14].

At the same time, industry data argues that the current land share used for bioplastics is small at a global scale. This view suggests that competition with food is limited today, but it still depends on how fast the sector scales and what feedstocks are chosen. Public reports underline that scale and location matter. If bio-based plastics expand using food crops, land pressure can rise and may compete with food production. These trends affect daily operations in several ways. Material scarcity raises input prices and price swings. It increases lead times. It makes qualification and quality control harder, especially in additive manufacturing, where feedstock must meet tight specifications. It also slows the adoption of new sustainable designs when a firm cannot secure a stable supply. The IEA's risk analysis shows that several minerals face high risk scores, with different exposure types such as supply risk and geopolitical risk [15, 16].

7.2 ADDRESSING OBSTACLES AND POTENTIAL SOLUTIONS FOR STRONGER OPERATIONAL SUSTAINABILITY

The most useful solutions reduce dependence on scarce inputs and make material loops stronger. They also improve transparency, so firms can measure progress and avoid false claims. In practice, the work needs to start from design and extend through process control, supply chain choices, and end-of-life systems.

A circular economy approach is central here. A widely used definition describes the circular economy as a system that keeps products and materials in use through maintenance, reuse, repair, refurbishment, remanufacture, recycling, and composting. This is not only an end-of-life idea. It is a design and business planning idea [17].

Additive manufacturing can support circular goals when it is used in the right way. It can reduce scrap in some part shapes. It can also enable on-demand production and local supply, which may reduce inventory waste and transport. But the net benefit depends on the full life cycle. LCA reviews of additive manufacturing stress that energy use, post-processing, and material choice can change the results. They also stress that results vary by process type and product use phase [18, 19].

Closed-loop production is a practical target for both polymers and metals. In polymer printing, recycling failed prints and support material into new filament is possible in many cases. Yet quality control is needed. Studies of recycled polymer printing highlight problems such as reduced strength after repeated recycling cycles and variability in flow and bonding. This means a closed loop should include testing of filament diameter, melt behaviour, and part strength [20].

For metals, powder reuse can reduce cost and waste. But it needs strict control. Research on powder reuse shows that reuse can change powder characteristics, which can affect part quality. Reviews also point to practical issues in recycling and reusing powder, such as contamination risk and changes in particle shape or oxygen pickup [21].

It is also important to look upstream. Metal powder production itself can carry a large footprint. A 2025 study on gas atomisation reports that powder production can have a climate impact of the same order as the printing step for powder-based processes. This means that sustainability work should cover both the printing machine and the supply chain for feedstock [22].

Recycling infrastructure needs scale, investment, and rules that reward recovery. Battery recycling is a strong example. The IEA shows that larger end-of-life volumes and higher collection can reduce demand for new mining over time. But it also shows that supply chains will still need a large new supply even with recycling growth [23].

Furthermore, experts emphasise that, despite the sustainability benefits of additive manufacturing compared to traditional methods, successfully using recycled materials presents a significant challenge. This creates a contradictory issue within circular economy models [24].

To mitigate this, strategic pathways are being developed, including the creation of closed-loop recycling systems and the design of biodegradable additives, which are essential for enhancing environmental outcomes in additive manufacturing [25]. One promising avenue for increasing the sustainability of additive manufacturing involves enhancing the reusability of materials, as evidenced by the high percentage of unused metal powder that can be filtered and reused directly in metal additive manufacturing [26]. Using existing materials significantly reduces waste and improves the economic viability of additive manufacturing processes [27, 28].

Public policy can support this shift. The EU sets 2030 benchmarks in its Critical Raw Materials Act approach. These include 10% extraction, 40% processing, and 25% recycling of the EU's annual needs for strategic raw materials. The same approach also aims to limit dependence on any single third country to 65% for each strategic raw material at key stages [29].

Even outside the EU, these targets signal a wider trend. Governments want stronger domestic and regional capacity for extraction, processing, and recycling to reduce supply shocks.

Digital tools can make sustainability work more precise. Life cycle assessment (LCA) is one of the most widely used tools to compare options and find 'hotspots' in the life cycle. The standard ISO 14040 sets the basic framework for LCA, including goal and scope definition, inventory, impact assessment, and interpretation [30].

For operational sustainability, this matters because it moves decisions away from guesswork and towards tracked data. It also helps avoid shifting harm from one stage to another.

Sensors and connected systems support day-to-day control. Industrial Internet of Things (IIoT) research shows how sensor networks and connected devices allow monitoring of equipment and production conditions. This can support better energy management, predictive maintenance, and lower scrap rates [31].

Digital twins can push this further. A digital twin is a live digital model of a physical system that is updated with real data. Research on digital twins in manufacturing describes how they can support sustainability and circular goals, mainly through better monitoring and better decision support [32].

Economic barriers remain a major block. Peer-reviewed work on sustainable manufacturing barriers shows that higher prices of sustainable products and limited supplier commitment are key problems. It also highlights a lack of access to information and training. These findings matter because they point to what firms need in practice: reliable suppliers, stable standards, and skills support. Financial incentives and procurement policies can help close the cost gap. They can also reduce risk for early adopters. In parallel, supply chain due diligence is becoming a basic expectation. The guidance on responsible mineral supply chains sets out a step-by-step approach for due diligence and risk-based management. It aims to help firms avoid links to serious harms and illegal activity in mineral supply chains.

Collaboration matters because no single firm can fix the whole chain. Universities can develop new materials and test methods. Firms can scale and qualify them. Policymakers can set rules and create markets for recycled content, repair, and safe disposal. This kind of joint work is already visible in the way standards are being built for sustainability measurement.

Skills and training are also essential. Sustainable manufacturing depends on people who can run new processes, use digital tools, and meet sustainability standards. The report highlights the need for 'skills for sustainability' in global supply chains. It links skills to compliance, resilience, and fair production.

In additive manufacturing, formal skill standards also exist. ISO/ASTM qualification standards for coordination personnel show that training and role clarity are now part of industrial quality systems. Finally, clear standards and certification systems reduce confusion and improve trust. Standards for LCA (ISO 14040) and ecolabelling (ISO 14024) support comparable claims. ISO 14024 sets principles for Type I environmental labels, where a third party awards the label based on multiple criteria and life cycle thinking. Certification programmes can also guide product choices. These programmes are described as voluntary third party certifications that recognise product performance across multiple sustainability categories using life cycle thinking. However, a significant challenge remains in the absence of standardised metrics for assessing material sustainability, hindering widespread adoption and integration into industrial processes [33]. This gap underscores the necessity for developing comprehensive and universally accepted sustainability assessment tools and methodologies to foster comparability and drive informed decision-making within the sector [34] (Table 7.1).

A major obstacle to achieving these goals is the lack of information sharing across the entire value chain. Moreover, there is a general lack of understanding of how to implement green manufacturing practices across different areas [35, 36]. The financial difficulties of adopting advanced sustainable technologies, combined with the limited resources of small and medium-sized businesses, worsen the situation [37, 38].

To overcome these obstacles, a collaborative approach is essential, emphasising cooperation among all stakeholders to integrate sustainable practices throughout the manufacturing supply chain [39]. This collaborative effort involves creating standardised methods and tools to ensure accurate information about processes. This approach aims to connect academic research with practical use [40].

In addition, overcoming the challenges identified requires a concerted effort to improve the skills of the workforce, enabling them to effectively use and manage

TABLE 7.1
Challenges and future directions in sustainable manufacturing.

Area	Key Challenge	Impact on Operations	Future Direction / Solution
Material Availability	Scarcity of critical metals and polymers	Increased cost and supply risk	Use recycled materials and alternative feedstocks
Resource Dependence	High reliance on virgin raw materials	Environmental damage and instability	Adopt circular economy and closed-loop systems
Recycling Limitations	Poor recyclability of composites and polymers	Material loss and waste generation	Improve material design for reuse and recycling
Energy Use	High energy demand in manufacturing processes	Higher emissions and operating costs	Shift to energy-efficient equipment and processes
Supply Chain	Concentrated and fragile supply chains	Production delays and uncertainty	Diversify suppliers and localise production
Technology Adoption	Slow adoption of sustainable technologies	Limited improvement in sustainability	Promote digital tools and process optimisation
Cost Factors	High initial cost of sustainable solutions	Hesitation in implementation	Focus on long-term cost benefits and policy support
Skills and Knowledge	Lack of a trained workforce	Inefficient use of new technologies	Invest in training and skill development
Policy and Standards	Lack of clear sustainability standards	Inconsistent implementation	Develop common guidelines and certifications

these new systems. This must be combined with the creation of strong policy frameworks that encourage sustainable practices and facilitate the transfer of technology across different industries [41, 42]. This comprehensive approach is crucial for successfully navigating the complexities of sustainable manufacturing and fostering an environment where economic success and environmental responsibility can thrive [43]. A concerted approach is essential to surmount these obstacles. This necessitates governmental interventions, encompassing incentives, tax reductions, and subsidies, to mitigate initial expenditures. Simultaneously, it is crucial to highlight the enduring advantages of diminished overall costs and enhanced energy efficiency [44].

Furthermore, addressing the main obstacles to sustainable manufacturing often requires tackling issues related to top management commitment, leadership skills, and technical expertise, all of which are crucial for progress [45]. This is especially relevant for small and medium-sized businesses, where a lack of understanding, knowledge, and government support makes it harder to implement sustainable manufacturing practices [46, 47].

REFERENCES

1. R. da S. Manca, R. Galeano, K. L. Sanches, A. A. do N. Sousa, A. L. Mezzomo, A. P. da Silva, J. B. P. Cavalcante, T. Deitos, A. L. G. de Carvalho, L. H. P. da Silva, S. D. de Morais, & R. C. M. Rodrigues, Additive manufacturing in industry 4.0 and its implications for sustainability in production processes. *Revista de Gestão Social e Ambiental*, 18 (2024) e06733. https://doi.org/10.24857/rgsa.v18n1-139.
2. A. Peace, A. Ramírez, M. L. M. Broeren, N. Coleman, I. Chaput, T. Rydberg, & G.-N. Sauvion, Everyday industry – Pragmatic approaches for integrating sustainability into industry decision making. *Sustainable Production and Consumption*, 13 (2017) 93. https://doi.org/10.1016/j.spc.2017.08.003.
3. P. Saxena, P. Stavropoulos, J. D. Kechagias, & K. Salonitis, Sustainability assessment for manufacturing operations. *Energies*, 13 (2020) 2730. https://doi.org/10.3390/en13112730.
4. K. Armstrong, G. Avery, A. Bhatt, R. Burton, A. Carpenter, J. Cresko, C. Dollinger, W. Dong, C. O. Iloeje, D. Graziano, J. Greenblatt, C. McMillan, W. R. Morrow, S. Nicholson, S. Nimbalkar, D. Steward, S. D. Supekar, K. Thirumaran, & S. Upasani, *Sustainable Manufacturing and the Circular Economy* (U.S. Department of Energy, Office of Scientific and Technical Information, 2023). https://doi.org/10.2172/1963668.
5. I. E. Agency, *Global EV Outlook 2023* (Organization for Economic Cooperation and Development, 2023). https://doi.org/10.1787/cbe724e8-en.
6. B. Zhou, Z. Li, Y. Zhao, C. Zhang, & Y. Wei, Rare earth elements supply vs. clean energy technologies: New problems to be solve. *Gospodarka Surowcami Mineralnymi–Mineral Resources Management*, 32 (2016) 29. https://doi.org/10.1515/gospo-2016-0039.
7. K.-H. Yeh, & G. Ni, Critical mineral supply chain security and multi-layered regional governance: Insights from Central Asia. *Trames Journal of the Humanities and Social Sciences*, 29 (2025) 271. https://doi.org/10.3176/tr.2025.3.04.
8. U. S. G. Survey, *Mineral Commodity Summaries 2025* (U.S. Department of the Interior, U.S. Geological Survey, 2025). https://doi.org/10.3133/mcs2025.
9. A. Leader, G. Gaustad, & C. W. Babbitt, The effect of critical material prices on the competitiveness of clean energy technologies. *Materials for Renewable and Sustainable Energy*, 8 (2019) 8. https://doi.org/10.1007/s40243-019-0146-z.
10. F. Creutzig, S. G. Simões, S. Leipold, P. Berrill, I. Azevedo, O. Y. Edelenbosch, T. Fishman, H. Haberl, E. G. Hertwich, V. Krey, A. T. Lima, T. Makov, A. Mastrucci, N. Milojevic-Dupont, F. Nachtigall, S. Pauliuk, M. Silva, E. Verdolini, D. P. van Vuuren, F. Wagner, D. Wiedenhofer, & C. Wilson, Demand-side strategies key for mitigating material impacts of energy transitions. *Nature Climate Change*, 14 (2024) 561. https://doi.org/10.1038/s41558-024-02016-z.
11. C. Xu, Q. Dai, L. Gaines, M. Hu, A. Tukker, & B. Steubing, Future material demand for automotive lithium-based batteries. *Communications Materials*, 1 (2020) 99. https://doi.org/10.1038/s43246-020-00095-x.
12. J. Wesselkämper, L. Dahrendorf, L. Mauler, S. F. Lux, & S. von Delft, A battery value chain independent of primary raw materials: Towards circularity in China, Europe and

the US. *Resources Conservation and Recycling*, 201 (2023) 107218. https://doi.org/10.1016/j.resconrec.2023.107218.

13. Y. Dong, H. Li, L. Wang, J. Chen, V. I. Sucală, & J. Jiang, Review on recycled materials for additive manufacturing. *International Journal of Precision Engineering and Manufacturing-Green Technology* (2025). https://doi.org/10.1007/s40684-025-00788-z.
14. Y. van der Meer, Sustainable bio-based materials: Opportunities and challenges, *Proceedings of the Biotech France 2017 International Conference*, Paris, France, June 28–30, (2017) 1–5. https://doi.org/10.26799/cp-biotechfrance2017.1.
15. Y. Lapko, P. Trucco, & C. Nuur, The business perspective on materials criticality: Evidence from manufacturers. *Resources Policy*, 50 (2016) 93. https://doi.org/10.1016/j.resourpol.2016.09.001.
16. I. E. Agency, *The Role of Critical Minerals in Clean Energy Transitions* (Organization for Economic Cooperation and Development, 2021). https://doi.org/10.1787/f262b91c-en.
17. W. R. Stahel, The circular economy. *Nature*, 531 (2016) 435. https://doi.org/10.1038/531435a.
18. V. Pusateri, S. I. Olsen, S. Kara, & M. Z. Hauschild, Life cycle assessment of metal additive manufacturing: A systematic literature review. *Research Portal Denmark*, 8 (2021) 1.
19. M. Gopal, H. G. Lemu, & E. M. Gutema, Sustainable additive manufacturing and environmental implications: Literature review. *Sustainability*, 15 (2022) 504. https://doi.org/10.3390/su15010504.
20. A. Patti, Assessment of recycled PLA-based filament for 3D printing. *MDPI (MDPI AG)*, 7(1) (2021) 16. https://doi.org/10.3390/iocps2021-11209.
21. L. Cordova, M. Campos, & T. Tinga, Revealing the effects of powder reuse for selective laser melting by powder characterization. *JOM*, 71 (2019) 1062. https://doi.org/10.1007/s11837-018-3305-2.
22. D. Ochs, K. K. Wehnert, J. Hartmann, A. Schiffler, & J. Schmitt, Sustainable aspects of a metal printing process chain with laser powder bed fusion (LPBF). *Procedia CIRP*, 98 (2021) 613. https://doi.org/10.1016/j.procir.2021.01.163.
23. E. Lorang, The resource, waste and climate conundrum: Essays on the economics of recycling. *Economics and Finance* (Université de Lorraine, NNT: 2022LORR0016, 2022).
24. J. Priyadarshini, R. K. Singh, R. Mishra, Q. He, & A. Braganza, Implementation of additive manufacturing in the healthcare supply chain for circular economy goals: Paradoxical tensions and solutions from an industry 5.0 perspective. *Information Systems Frontiers* (2024). https://doi.org/10.1007/s10796-024-10482-1.
25. V. Phutela, & S. Juneja, Innovative approaches to recycling and waste reduction in 3D printing. *International Journal of Environmental Science and Technology*, 22 (2025) 17551. https://doi.org/10.1007/s13762-025-06722-9.
26. M. Despeisse, M. Baumers, P. Brown, F. Charnley, S. Ford, A. Garmulewicz, S. Knowles, T. Minshall, L. Mortara, F. Reed-Tsochas, & J. A. Rowley, Unlocking value for a circular economy through 3D printing: A research agenda. *Technological Forecasting and Social Change*, 115 (2016) 75. https://doi.org/10.1016/j.techfore.2016.09.021.
27. H. Hegab, N. Khanna, N. Monib, & A. Salem, Design for sustainable additive manufacturing: A review. *Sustainable Materials and Technologies*, 35 (2023) e00576. https://doi.org/10.1016/j.susmat.2023.e00576.
28. B. Ferreira, A. A. de Campos, R. Casati, A. Gonçalves, M. Leite, & I. Ribeiro, Technological capabilities and sustainability aspects of metal additive manufacturing. *Progress in Additive Manufacturing*, 9 (2023) 1737. https://doi.org/10.1007/s40964-023-00534-4.

29. E. Righetti, & V. Rizos, The EU's quest for strategic raw materials: What role for mining and recycling? *Intereconomics*, 58 (2023) 69. https://doi.org/10.2478/ie-2023-0015.
30. M. Sutton, Fundamental tools for managing sustainability. *RSC Sustainability*, 1 (2023) 1311. https://doi.org/10.1039/d3su90030d.
31. X. Chen, M. Despeisse, & B. Johansson, Environmental sustainability of digitalization in manufacturing: A review. *Sustainability*, 12 (2020) 10298. https://doi.org/10.3390/su122410298.
32. J. Mügge, A. Seegrün, T.-K. Hoyer, T. Riedelsheimer, & K. Lindow, Digital twins within the circular economy: Literature review and concept presentation. *Sustainability*, 16 (2024) 2748. https://doi.org/10.3390/su16072748.
33. J. M. Schoenung, O. A. Ogunseitan, & L. Heine, Sustainability-informed materials selection, design, discovery, and development. *MRS Bulletin*, 50 (2025) 1225–1242. https://doi.org/10.1557/s43577-025-00935-6.
34. E. Hansen, J. Korhonen, R. Panwar, & M. Hakovirta, The quandary of sustainability-oriented innovations. In D. D'Amato, A. Toppinen, & R. Kozak (Eds.), *Routledge eBooks* (Informa, 2022), p. 185. https://doi.org/10.4324/9781003003588-15.
35. O. Klymenko, L. L. Halse, & B. Jæger, Towards sustainability: The manufacturers' perspective. In B. Lalic, V. Majstorovic, U. Marjanovic, G. von Cieminski, & D. Romero (Eds.), *IFIP Advances in Information and Communication Technology* (Springer Science+Business Media, 2020), p. 174. https://doi.org/10.1007/978-3-030-57997-5_21.
36. D. Kannan, K. M. Shankar, & P. Gholipour, Paving the way for a green transition through mitigation of green manufacturing challenges: A systematic literature review. *Journal of Cleaner Production*, 368 (2022) 132578. https://doi.org/10.1016/j.jclepro.2022.132578.
37. J. K. Runtuk, P. K. Ng, & O. S. Yin, Challenges and solutions in working with green suppliers: Perspective from a manufacturing industry. *Sustainability*, 16 (2024) 8744. https://doi.org/10.3390/su16208744.
38. R. Chourasiya, S. Pandey, R. K. Malviya, & A. A. Pujara, Towards sustainable success: A framework for assessing performance of sustainable manufacturing adoption in Indian textile industry. *Sustainable Futures*, 7 (2024) 100216. https://doi.org/10.1016/j.sftr.2024.100216.
39. M. A. Rhaman, & H. Ahmed, Innovative approaches to sustainable supply chain management in the manufacturing industry: A systematic literature review. *Deleted Journal*, 3 (2024) 1. https://doi.org/10.62304/jieet.v3i02.81.
40. M. M. Smullin, K. R. Haapala, M. Mani, & K. C. Morris, Using industry focus groups and literature review to identify challenges in sustainable assessment theory and practice. *Proceedings of the ASME 2016 International Design Engineering Technical Conferences and Computers and Information in Engineering Conference.* Vol 4: 21st Design for Manufacturing and the Life Cycle Conference; 10th International Conference on Micro- and Nanosystems. Charlotte, North Carolina, USA. August 21–24 (2016). https://doi.org/10.1115/detc2016-60216.
41. A. Ö. Rönnbäck, L. Larsson, K. Johansen, U. Harlin, & K. Säfsten, Resilient and sustainable production systems: Towards a research agenda. *International Journal of Manufacturing Research*, 19 (2023) 343–365. https://doi.org/10.1504/ijmr.2024.10058315.
42. K. Challouf, A. Alhloul, & N. Németh, Mapping the role of industry 4.0 technologies in green supply chain management: A bibliometric and structured text analysis. *Discover Sustainability*, 6 (2025) 949. https://doi.org/10.1007/s43621-025-01827-0.

43. D. Kannan, P. Gholipour, & C. Bai, Smart manufacturing as a strategic tool to mitigate sustainable manufacturing challenges: A case approach. *Annals of Operations Research*, 331 (2023) 543. https://doi.org/10.1007/s10479-023-05472-6.
44. S. Tahmasseby, S. Salam, & M. R. Maleki, Sustainable construction and life cycle assessment in Qatar: Industry challenges, adoption, and policy implications. *Environment Development and Sustainability* (2025). https://doi.org/10.1007/s10668-025-06661-y.
45. M. C. Mandal, N. Mondal, & A. Ray, Analyzing and optimization of the critical barriers of sustainable manufacturing. *Process Integration and Optimization for Sustainability*, 9(3) (2025) 1125–1143. https://doi.org/10.1007/s41660-025-00493-z.
46. C. Alayón, K. Säfsten, & G. Johansson, Barriers and enablers for the adoption of sustainable manufacturing by manufacturing SMEs. *Sustainability*, 14 (2022) 2364. https://doi.org/10.3390/su14042364.
47. W. S. Yip, H. Zhou, & S. To, A critical analysis on the triple bottom line of sustainable manufacturing: Key findings and implications. *Environmental Science and Pollution Research*, 30 (2023) 41388. https://doi.org/10.1007/s11356-022-25122-x.

8 Conclusions

This work has examined operational sustainability in manufacturing with a strong focus on material selection, product design, and additive manufacturing processes. The study shows that sustainability cannot be not achieved through a single decision or technology. It emerges from coordinated actions taken across the entire product life cycle, from material sourcing to end-of-life management.

The discussion in Chapter 1 established that operational sustainability is a practical and measurable concept. It extends beyond environmental concerns and includes economic performance and social responsibility. Integrating sustainability into daily manufacturing decisions enables industries to reduce waste, manage resources efficiently, and remain competitive in the long term. The triple bottom line framework provides a useful foundation for evaluating such decisions.

Chapter 2 highlighted that material selection is one of the most influential factors in sustainable manufacturing. The choice of materials directly affects energy consumption, emissions, recyclability, and product lifespan. Life cycle assessment was identified as a key tool for comparing materials on a scientific basis rather than assumptions. The chapter showed that sustainable materials, such as recyclable metals, bio-based polymers, and recycled feedstocks, can significantly lower environmental impacts when used with proper design and process control.

The role of product design, discussed across multiple chapters, was shown to be central to sustainability outcomes. Design strategies such as lightweighting, topology optimisation, and part consolidation reduce material usage without compromising performance. Additive manufacturing offers unique advantages in this area by enabling complex geometries and customised designs that are difficult to achieve using conventional methods.

Chapters 3 and 4 demonstrated that additive manufacturing is not automatically sustainable. Energy consumption, machine utilisation, build planning, and post-processing play a major role in determining the overall environmental footprint. Case-based evidence showed that non-productive phases, such as machine warm-up and cool-down, can account for a large share of total energy use. Therefore, equipment selection, build volume optimisation, and operational planning are critical for improving energy efficiency.

DOI: 10.1201/9781003633754-8

Chapter 5 explored novel approaches and emerging materials that support operational sustainability. Bio-based polymers, recycled feedstocks, sustainable metal powders, and hybrid material systems were discussed as promising pathways. However, the chapter also highlighted practical limitations, such as material variability, recyclability challenges, and quality control issues. These findings confirm that sustainability benefits depend on controlled implementation rather than material labels alone.

Real-world validation was provided in Chapter 6, which presented industrial case studies across multiple sectors. These examples showed that sustainability initiatives can deliver measurable benefits, including reduced energy use, lower emissions, cost savings, and improved productivity. The integration of Industry 4.0 technologies, such as Internet of Things (IoT), digital twins, and data analytics, was shown to enhance transparency and enable continuous sustainability monitoring. Successful cases demonstrated that sustainability and profitability are not conflicting goals but can reinforce each other.

Chapter 7 addressed the challenges and future directions for operational sustainability. Material scarcity, supply chain concentration, recycling limitations, and energy-intensive feedstock production were identified as major risks. The analysis showed that future progress depends on strengthening circular economy practices, improving recycling infrastructure, and designing products for multiple life cycles. Additive manufacturing can support these goals, but only when supported by standards, quality control, and system-level thinking.

Overall, this work concludes that operational sustainability in manufacturing requires an integrated approach. Sustainable outcomes are achieved when material selection, design decisions, process optimisation, and digital technologies are aligned. Additive manufacturing offers powerful tools to support this transition, but its benefits depend on informed operational choices. The findings of this study provide a structured framework for researchers, designers, and industry practitioners to implement sustainability in practical and measurable ways.

Future work should focus on improving data availability for life cycle assessment, developing robust recycling routes for advanced materials, and expanding the use of real-time monitoring systems. Strengthening collaboration between academia, industry, and policymakers will be essential to scale sustainable manufacturing practices and meet long-term environmental and economic goals.

Index

www.ingramcontent.com/pod-product-compliance
Ingram Content Group UK Ltd.
Pitfield, Milton Keynes, MK11 3LW, UK
UKHW020207200926
475702UK00004B/660
* 9 7 8 1 0 3 2 9 9 9 1 7 3 *